COLORADO ROCKS!

A Guide to Geologic Sites in the Centennial State

MAGDALENA S. DONAHUE
AND MARLI B. MILLER

2021
Mountain Press Publishing Company
Missoula, Montana

GEOLOGY ROCKS!

A state-by-state series that introduces readers to some of the most compelling and accessible geologic sites in each state.

© 2021 by Magdalena S. Donahue and Marli B. Miller
First Printing, October 2021
All rights reserved

Cover photo:
Painted Wall in Black Canyon of the Gunnison National Park. —Magdalena S. Donahue photo

Photographs by Magdalena S. Donahue and Marli B. Miller unless otherwise credited.

Illustrations constructed by Chelsea M. Feeney
(www.cmcfeeney.com)

Library of Congress Cataloging-in-Publication Data

Names: Donahue, Magdalena S., 1982- author. | Miller, Marli B., 1960-
   author.
Title: Colorado rocks! : a guide to geologic sites in the centennial state
   / Magdalena S. Donahue and Marli B. Miller.
Description: First. | Missoula, Montana : Mountain Press Publishing
   Company, 2021. | Series: Geology rocks! | Includes bibliographical
   references and index. | Summary: "To discover the geologic novelties of
   the Centennial State, all that is required is a good map, a sense of
   adventure, and *Colorado Rocks*, a guide to 50 of the most compelling
   geologic sites in Colorado." —Provided by publisher.

Identifiers: LCCN 2021031258 | ISBN 9780878427055 (paperback)
Subjects: LCSH: Geology—Colorado—Guidebooks. | Colorado—Guidebooks. |
   Mines and mineral resources—Colorado.
Classification: LCC QE91 .D66 2021 | DDC 557.88—dc23
LC record available at https://lccn.loc.gov/2021031258

PRINTED IN THE UNITED STATES

MP Mountain Press
PUBLISHING COMPANY
P.O. Box 2399 • Missoula, MT 59806 • 406-728-1900
800-234-5308 • info@mtnpress.com
www.mountain-press.com

PREFACE AND ACKNOWLEDGMENTS

The geology of Colorado is written in the rocks. From this great book are here presented a few translations of a few paragraphs. The scenery of Colorado is a gallery incomparable. Words lack form and light—the essence and soul of scenery. At best they can but call attention to the elements associated in the picture. They cannot convey the beauty and harmony of the assemblage.

—**RUSSELL D. GEORGE**, first director (1908–1926) of the Colorado Geological Survey

I started writing this book as I came close to finishing my dissertation. Having spent years of focused study on the evolution of the Rocky Mountain landscape, I wanted to shout my findings from the (literal) mountaintops. As I enthusiastically waxed poetic about the geology of the Rockies with friends and family on every hike or road trip we took, I realized I wanted to share the geologic information I had gathered with an even wider audience. Geology is the utmost of tangible and visual sciences; a book with abundant photographs and maps seemed the best way to share the visual and scientific aspects of Colorado with a general audience. The crafting of this book has been a light point through many life adventures during many years of work, study, and play.

Many thanks to my husband, John, for his willingness to spend our free time exploring the mountains and canyons of the Rockies, for supporting me in my long solo drives gathering pictures and ideas for this book, and for his constant creative and happy encouragement. Thanks also to my children, my two daughters and son, who knew their mom was going to be away for a couple of weeks once in a while to "go look at rocks" and who are now my trusty hiking and naturalist partners at nine, six, and one years of age. Thanks to Marli Miller who encouraged me to consider writing a book that combined my geologic and artistic training and who came to be a wonderful and practiced coauthor. Thanks to Amy Ellwein for her thorough and thoughtful review of the manuscript. Thanks to Jenn Carey and Mountain Press for walking me through the first stumbling, hesitant attempts at writing a book.

This book discusses geologic locations and phenomena that are largely accessible via paved road. A few of these locations, however, are very remote; accessing them requires advance planning, current maps in addition to GPS or digital/cellular navigational tools, and a vehicle capable of handling the occasional dirt road and snowy or icy conditions. While many of the locations discussed can be seen from the highway or from public parks, we suggest hiking to view some features and their surrounding areas. Most of the locations we discuss in this book are on public land. Collecting of rocks, plants, and other artifacts, as well as defacing rocks or trees with initials, is prohibited; please help preserve geological formations for future generations.

WESTERN CANYONS AND PLATEAUS CENTRAL ROCKIES FRONT RANGE AND GREAT PLAINS

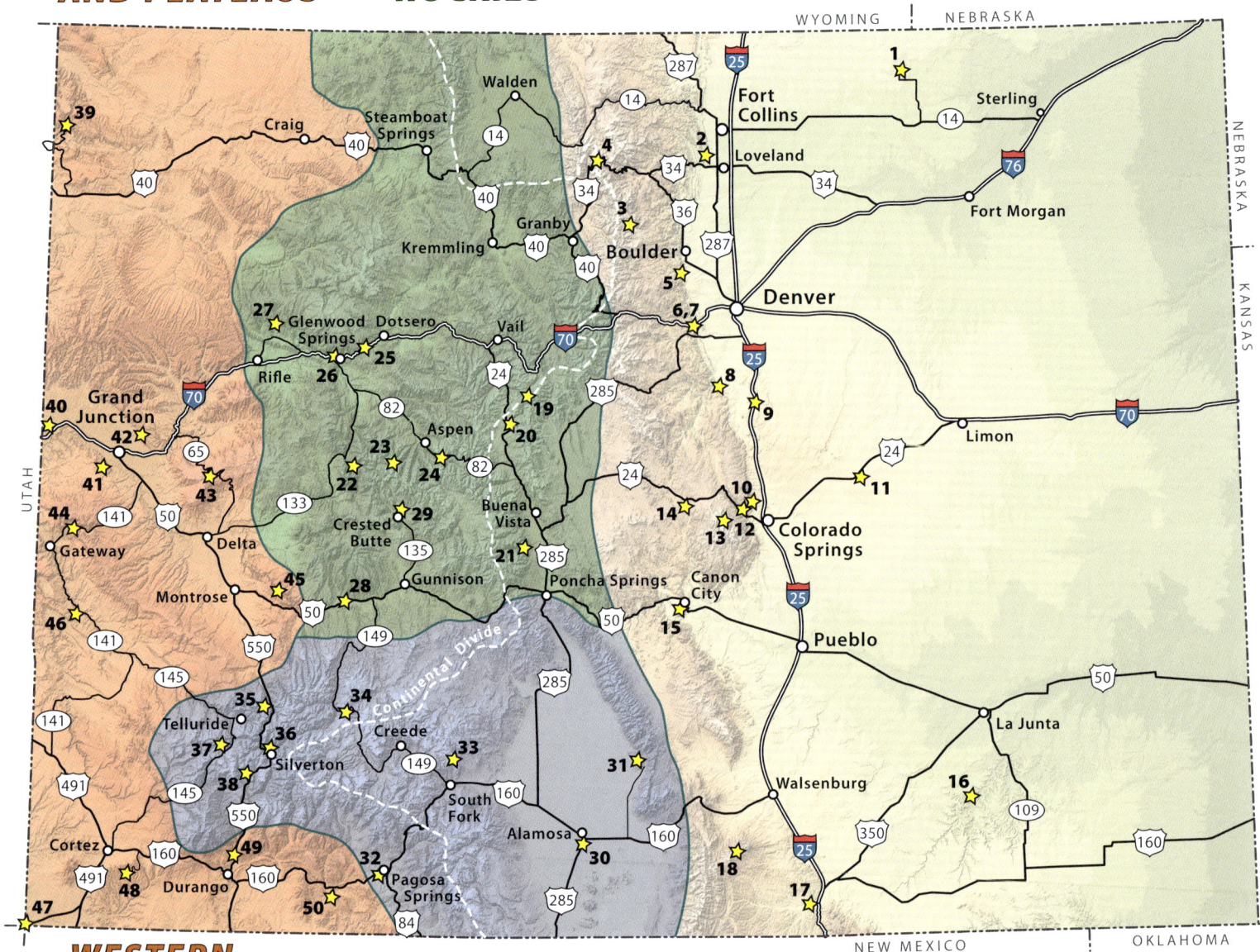

WYOMING NEBRASKA

Walden

Craig Steamboat Springs

39

1

Sterling

Fort Collins

4 2 Loveland

Fort Morgan

3 Boulder

Kremmling Granby

Dotsero Vail

27 Glenwood Springs

25

Rifle 26

5

6,7 Denver

8

9

19

20

40 Grand Junction

42

41

43

23 Aspen

22 24

Limon

11

24

285

70

10 Colorado Springs

14

13 12

Crested Butte 29

21 Poncha Springs

Buena Vista

Gateway 44

46

Montrose 45 28 Gunnison

Canon City

15

Delta

Pueblo

Continental Divide

35 34

Telluride 36 Creede 33

37 31

La Junta

Silverton 149

38 South Fork 16

Cortez 49 160 Alamosa 30 Walsenburg

48 Durango 32 Pagosa Springs 18 350

47 50 84 285 17 160

NEW MEXICO OKLAHOMA

WESTERN CANYONS AND PLATEAUS

SAN JUAN MOUNTAINS AND RIO GRANDE RIFT

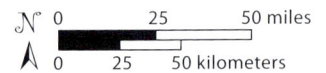

N 0 25 50 miles
0 25 50 kilometers

UTAH NEBRASKA KANSAS

CONTENTS

ERA	PERIOD	EPOCH	AGE	IMPORTANT GEOLOGIC EVENTS IN COLORADO (site locations in parentheses)	KEY ROCK UNITS
CENOZOIC	QUATERNARY	HOLOCENE	0.01	Continued erosion of high topography shapes the modern landscape (31, 34, 44, 45); continued rise of groundwater at hot springs and depositing minerals (26, 27, 32); filling and usage of aquifers (1, 30)	alluvium gravels sand dunes glacial till
CENOZOIC	QUATERNARY	PLEISTOCENE	2.6	Major glaciations shape the land (3, 4, 24, 37)	
CENOZOIC	TERTIARY / NEOGENE	PLIOCENE	5.3	Basalt erupts 10 million years ago, caps mesas today (43); recent uplift triggers downcutting (39)	Ogallala Formation White River Formation Castle Rock Conglomerate Green River Formation Denver Formation Dawson Arkose Fish Canyon Tuff Wall Mountain Tuff
CENOZOIC	TERTIARY / NEOGENE	MIOCENE	23	Rio Grande rift extension begins during Oligocene to Miocene time (21, 30)	
CENOZOIC	TERTIARY / PALEOGENE	OLIGOCENE	33.9	San Juan and 39-Mile volcanism, 38 to 23 million years ago, bring major accumulation of ash and tuff to CO (9, 14, 21, 28, 33, 36, 38); laccoliths intrude (29)	
CENOZOIC	TERTIARY / PALEOGENE	EOCENE	56		
CENOZOIC	TERTIARY / PALEOGENE	PALEOCENE	66	K-Pg boundary: asteroid impact, global extinctions (17)	
MESOZOIC	CRETACEOUS			The Laramide orogeny begins 75 million years ago, ushering out the last seas in CO; brings CO to near-modern elevations; initiates major canyon-carving; deposits ore (2, 6, 15, 19, 20, 25, 36)	Mt. Princeton batholith Laramie Formation Fox Hills Sandstone Pierre Shale Niobrara Formation Mesaverde Group Mancos Shale Dakota Group
MESOZOIC	CRETACEOUS		145	The Western Interior Seaway shallows and deepens, bringing swampy, then marine conditions and deposition in a tropical climate throughout much of the Cretaceous (2, 7, 11, 42, 47, 48, 50)	
MESOZOIC	JURASSIC		201	Abundant dinosaurs roam largely arid landscapes (7, 16, 40) Sand dune, swamps, and rivers deposit a range of sedimentary units (40, 41, 46) Uranium and vanadium are deposited at this time	Morrison Formation Ralston Creek Formation Navajo Sandstone Entrada Sandstone Wingate Sandstone
MESOZOIC	TRIASSIC		252	The Ancestral Rockies are largely eroded and dinosaurs are present in abundance; these animals roam arid plains and sand dune fields, and are preserved in red bed units	Chinle Formation Lykins Formation
PALEOZOIC	PERMIAN		299	The Ancestral Rockies are eroding (6, 8, 10, 23), forming deep basins filled with thick red bed sedimentary deposits; early reptile tracks are preserved in the Lyons Formation	Maroon Formation Lyons Formation
PALEOZOIC	PENNSYLVANIAN		323	The Ancestral Rockies orogeny begins; older overlying sediments are eroded from uplifts (5)	Fountain Formation Hermosa Formation Minturn Formation
PALEOZOIC	MISSISSIPPIAN		359	Shallow seas deposit thick layers of limestone; the seas then retreat (20, 22, 25, 26, 27)	Leadville Limestone Molas Formation
PALEOZOIC	DEVONIAN		419	The Colorado region is covered in shallow seas, depositing limestone	Dyer Dolomite Ouray Limestone Elbert Sandstone
PALEOZOIC	SILURIAN		444	The Colorado region is covered in shallow seas; a period of erosion removes most Silurian rock	
PALEOZOIC	ORDOVICIAN		485	The Colorado region is covered in deep ocean, depositing limestone and dolomite (12)	Manitou Limestone Harding Sandstone Fremont Dolomite
PALEOZOIC	CAMBRIAN		541	A sea floods the Colorado region; the first life is preserved in trilobites and shelled animals	Sawatch Sandstone
PRECAMBRIAN	PROTEROZOIC EON		2,500	A long period of erosion levels the land: the flat surface becomes the Great Unconformity (7, 25, 35, 49) The Uinta Mountain Group deposited sometime after 766 million years ago. Intrusion of granitic materials into crust at 1.4 (3) and 1.08 billion years ago (13) The Colorado orogeny from 1.78 to 1.65 billion years ago resulted in continental and island arc collision, accretion, building of new continental basement rock The oldest continental sedimentary rocks are deposited (15, 45)	Uinta Mountain Group Pikes Peak Granite Silver Plume Granite Boulder Creek Granite Idaho Springs Formation Vernal Mesa Quartzite Eolus Granite Twilight Gneiss Uncompahgre Formation
PRECAMBRIAN	ARCHEAN EON		3,850	Metamorphism is recorded at approximately 2.7 billion years ago in the Wyoming craton at the northern edge of Colorado	Red Creek Quartzite
PRECAMBRIAN	HADEAN EON		4,600	approximate age of Earth	

ages in millions of years before present

A BRIEF HISTORY OF COLORADO GEOLOGY

From the flat sedimentary cover of the Great Plains, across the volcanic and crystalline rocks of the Rocky Mountains, to the striking red rocks of the western plateaus, Colorado is home to some of the most spectacular and varied geology in the world. Colorado's geologic history extends over 2.7 billion years and includes continental collisions and mountain building, massive caldera-forming volcanoes, inundation by many different oceans and seaways, and erosional episodes that moved rocks, particle by particle, from the highest of peaks to the bottoms of seas. Recent uplift, downcutting rivers, and enormous ice age glaciers sculpted the state into the jaw-dropping landscapes we now know and love.

Not only does Colorado preserve the rocks of this long history, but the state also preserves a record of the life that occupied past landscapes and waters. From dinosaur tracks and bones to petrified tree stumps and imprints of leaves, fossils in the sedimentary rocks of Colorado give insight into past environments, depositional settings, and changing topography. Brachiopods, conodonts, gastropods, sponges, trilobites, and worms are some of the oldest fossils found in Colorado. These invertebrates inhabited a shallow sea in early Paleozoic time. Colorado also hosts three notable fossil-rich sedimentary units: the Morrison Formation, known for its Jurassic dinosaur bones, the Green River Formation with its fish, insects, and numerous tree fossils from Eocene time; and the White River Group, known for extensive mammal fossils including camels, elephants, mammoths, and early horses that lived from Eocene to Oligocene time. A spectacular location for Eocene-age fossils of plants, insects, and fish is Florissant Fossil Beds National Monument. These fossils have been critical to understanding past regional climates and depositional environments and may be important in understanding the timing of when the Rocky Mountains gained their elevation.

The Rocky Mountains, part of the larger North American Cordillera that extends more than 3,000 miles from Alaska to northern Mexico, tower above the plains to the east and plateaus to the west. Colorado boasts 53 mountain peaks that exceed 14,000 feet (4,270 meters) in elevation, known as Fourteeners. The high peaks of the Rockies give Colorado the highest mean elevation of any US state, and it is the only state to be entirely above 3,317 feet (1,000 meters) in elevation. The Continental Divide snakes through the state, and some of Colorado's highest peaks are considerably east of the divide because the Arkansas River managed to cut through the massive Front Range.

Colorado hosts the headwaters of several major rivers. Draining the west side of the Continental Divide are the

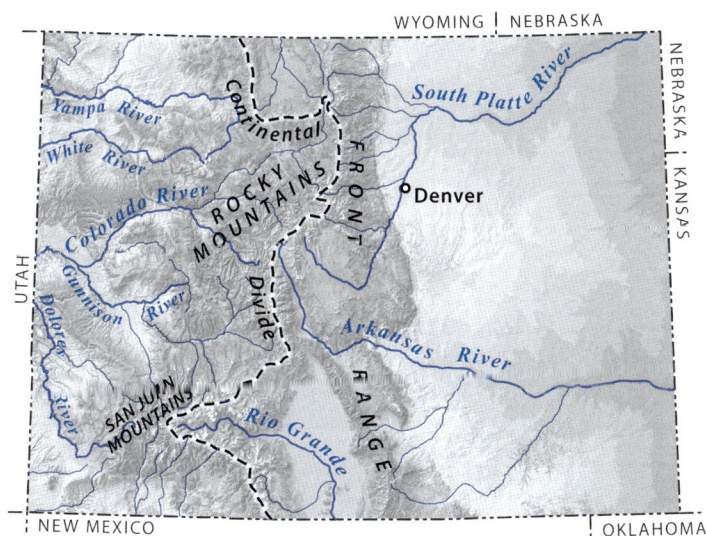

The Continental Divide snakes through Colorado, dividing the headwaters of major rivers.

1

The Gunnison River sliced through more than 2,000 feet of basement rock to form the spectacular Black Canyon. Steep sided, rugged, and spectacular, this gorge is one of many bedrock canyons carved into the Colorado high country by the incredible power of water. —Magdalena S. Donahue photo

Colorado, Gunnison, Yampa, and Dolores Rivers. On the east side of the divide are the Rio Grande, Arkansas, and South Platte Rivers. The erosive action of running water and the link these major rivers have to the sea have enabled the carving of deep and striking canyons into hard crystalline rock, notably the Black Canyon of the Gunnison, the Royal Gorge of the Arkansas, and Glenwood Canyon on the Colorado River. Geologists call the hard crystalline rock of these canyons and other places the "basement rock" because it's at the bottom and usually buried beneath younger sedimentary rock layers. However, basement rock can be present and exposed at high elevations when it is uplifted into mountains. Many of Colorado's peaks are composed of exposed basement rock.

During many times in Colorado's lengthy history, mountains formed in the state. Mountain building events are called orogenies and usually involve the movement of tectonic plates over the surface of Earth. Where plates meet, they often crumple and push the crust into mountains, melt rock to form magma, which can fuel volcanoes, and even subduct under one another. Tectonic plate movement far to the west along the margin of North America was one of the factors responsible for uplifting Colorado's modern Rocky Mountains.

Archean and Proterozoic Time

The oldest rocks in Colorado were part of the Wyoming craton, a continent that existed more than 2.5 billion years ago and whose remains are now centered on Wyoming and dip into far northwestern Colorado. We don't know when Colorado's oldest rocks first formed, but we know they were metamorphosed 2.7 billion years ago in Archean time.

The next-oldest rocks in Colorado formed during the very first assembly of the North American continent, a collision of tectonic plates between 1.78 to 1.65 billion years ago in Proterozoic time. Known as the Colorado orogeny, the mountain building event took place where oceanic crust along the edge of the Wyoming craton was subducting beneath an adjacent plate to the southeast. Volcanic islands formed, and magma intruded into the overriding plate along the subduction zone. Eventually these rocks were welded to the Wyoming craton along what geologists call a suture zone. Similar to how scars change the pattern of our skin, this plastering of segment after segment of oceanic volcanic islands onto the Wyoming craton resulted in a pervasive northeast-trending geologic texture, or series of scars, in the Earth's crust. This pattern of scars influenced many later geologic events and formations in Colorado. Collectively, these continental bodies, along with others to the north and east, became the embryonic North American continent.

Much happened to these rocks over the next 500 million years as the crust compressed and stretched. These old rocks record episodes of metamorphism, mountain building, and the breaking apart, or rifting, of the continent by tensional stress. Magmas intruded the continental rock about 1.4 billion years ago. To see the cooled remains of this ancient magma, just look to the granite of Longs Peak in Rocky Mountain National Park. More magma intruded about 1.08 billion years ago, the remains of which are present at Pikes Peak west of Colorado Springs.

Except for rocks in and near Dinosaur National Monument in the northeastern corner of the state, there is no direct geologic record of what transpired in Colorado after the magma of Pikes Peak intruded until about the beginning of Paleozoic time—from 1.08 billion years ago to 0.51 billion years ago. Nearly 500 million years is a long span of time to have no rock or deposit recording the geologic history. Rocks record information of how they formed and

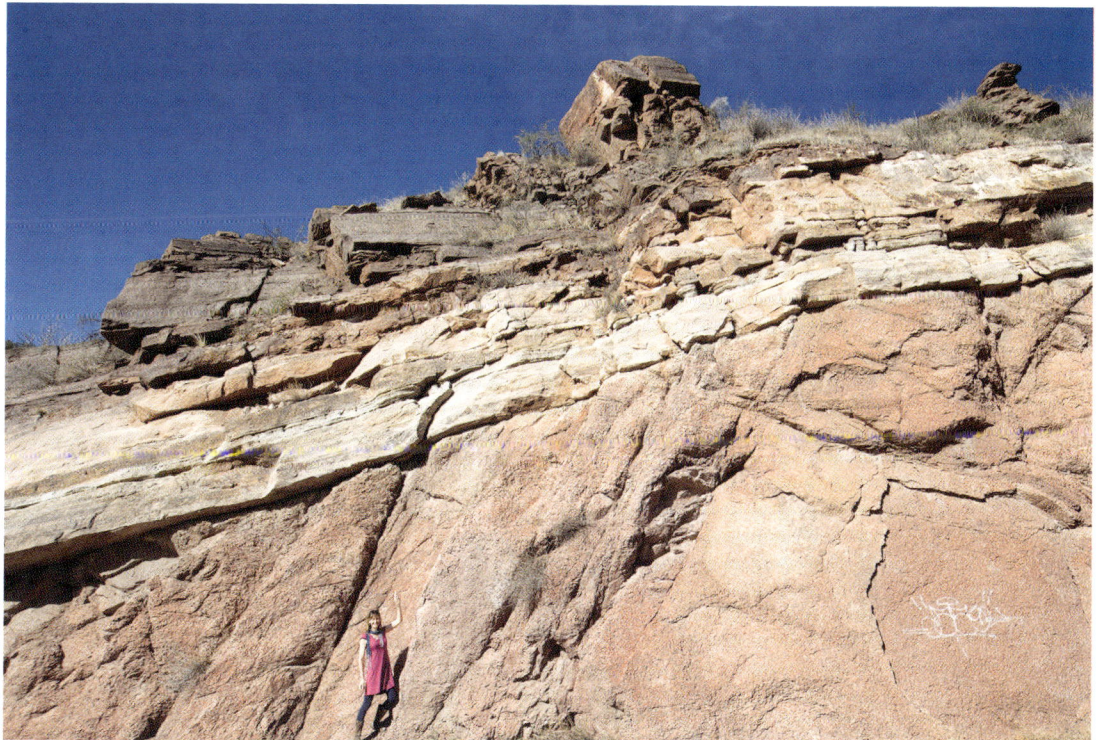

The Sawatch Sandstone of Cambrian age (tilted, layered rock at top) was deposited directly on an eroded surface of the Pikes Peak Granite (massive reddish rock at bottom) in an exposure of the Great Unconformity near Manitou Springs. —Marli B. Miller photo

when and what the surrounding ancient environments were like. When the rock record is present, geologists have a detailed look into the past. When it is missing, so too are the details.

Sediments were likely deposited during this time. However, just because a rock existed at one time does not guarantee its continued presence. Rocks are often eroded, leaving a gap in the rock record. Geologists call these gaps of missing rock *unconformities*. However, even these periods of missing rock give geologists information on what was going on at that time and place. There must have been a lot of erosion that removed all the evidence from this time!

This notable period of missing rock is known as the Great Unconformity and is found in many places across western North America. During this time of sustained erosion, most sedimentary rocks present were removed and transported to ocean basins. The basement rock was exposed at the surface and beveled into a flat plain. The Great Unconformity, famously noted in the Grand Canyon between the old Vishnu Schist at the bottom of the canyon and the younger Paleozoic layered rocks that you can see in the canyon walls, is also visible in multiple places in Colorado. Any trail that starts in the tilted sedimentary rock at the edge of the Front Range and climbs up into the granitic mountains crosses the Great Unconformity, although it often lies hidden, covered by young unconsolidated sediments and vegetation. The oldest sedimentary rock in Colorado above the Great Unconformity is Cambrian-age sandstone.

Paleozoic Time

Following the erosion that flattened the landscape into a plain of ancient crystalline rock, the North American continent was at low latitudes just north of the equator. Sands along the shore of a tropical sea are the first deposits preserved above the eroded surface of the unconformity. These beach sands became the Cambrian-age Sawatch Sandstone, which can be seen lying above Proterozoic basement rock on both sides of the Rocky Mountains. Sea levels gradually rose, and soon Colorado was covered in shallow tropical seawater. Shallow layers of fine-grained marine sediments were deposited in these quiet waters over the top of the sandstone on the continental shelf.

Ordovician time brought the accumulation of sediments that would become the Manitou Limestone, Harding Sandstone, and Fremont Dolomite. Deposition continued in Silurian time, but a short period of erosion removed much of that rock. Renewed deposition in shallow seas continued in Devonian time.

The only rock in Colorado from Silurian time is a piece of limestone found in an igneous kimberlite pipe in the State Line mining district on the Colorado-Wyoming border, and there are no other similar age rocks nearby. Kimberlites are most well-known for being the source of diamonds. These deep, narrow fissures are generally less than half a mile across at the surface and extend deep into the crust, narrowing to only perhaps 1 yard across at the very bottom. Thus, they take on the shape of a long, narrow carrot extending into the upper mantle. Diamonds form deep within the upper mantle, generally between 90 to 300 miles below the surface, and rocks containing diamonds only reach the surface when they are violently erupted vertically through the overlying crust, propelled by a gas-rich magma. Diamonds were found in the State Line district in 1976 at the Kelsey Lake Diamond Mine, which is no longer in operation.

In Mississippian time, carbonate deposition resulted in the Leadville Limestone, which was tectonically elevated above sea level shortly after deposition. As anyone knows who lives where limestone is at the surface, sinkholes and caves form in this rock type when it's exposed to surface water. Those features formed in parts of the Leadville Limestone before younger sediments were deposited above it. The Leadville unit has become famous in mining history because many precious ore metals were deposited in it much later by mineral replacement.

The uplift that began in the Mississippian Period continued into Pennsylvanian time, leaving a long series of depositional units tied to a retreating seaway: limestones, shales, and sandstones. By the end of the Pennsylvanian Period, marine deposition had ended in Colorado because a new mountain building event and its related uplift was underway.

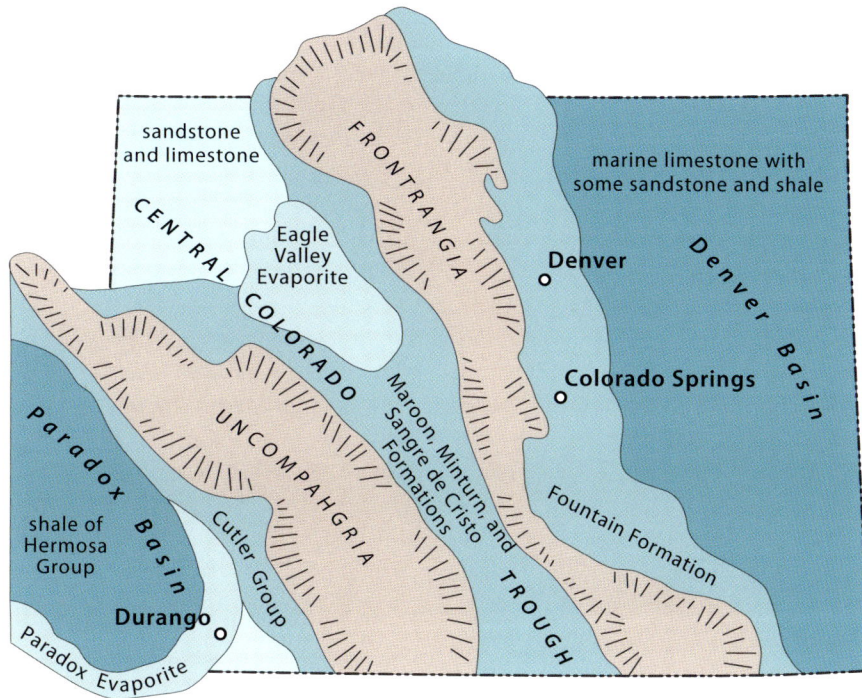

The Ancestral Rocky Mountain orogeny uplifted two mountain ranges that then proceeded to erode. —Modified from Williams and Chronic, 2014

The Ancestral Rocky Mountain orogeny, occurring about 300 million years ago, fundamentally reshaped the Late Pennsylvanian topography of Colorado. Two highlands, Frontrangia and Uncompahgria, formed when large blocks of Earth's crust were uplifted along faults. Overlying Paleozoic rock layers draped over the uplifted blocks and into the Central Colorado trough that separated the two highlands. The current idea as to what caused this mountain building is that North America was colliding with South America and Africa far to the south and east. The Appalachian Mountains formed at the collision zone, while farther west the crust buckled upward, forming the highlands of the Ancestral Rocky Mountains.

It might seem counterintuitive, but the primary evidence of this uplift are deposits composed of copious amounts of coarse-grained material eroded off these rising mountains. Rivers carried the sediment to basins at the edges of the uplifts. The Fountain and Lyons Formations, a series of red sandstones and conglomerates and dune and river sandstones, were deposited in the Denver Basin. The Minturn and Maroon Formations, a series of sandstones, conglomerates, and minor limestones and evaporites, were deposited in the Central Colorado trough. Sediments shed from the Uncompahgria uplift into the Paradox Basin include the Hermosa and Cutler Groups, famously red conglomerates, sandstones, and siltstones. Uplifted topography is subject to vigorous erosion, and by the end of the Permian Period, the Ancestral Rocky Mountains had been largely eroded away.

Mesozoic Time

During the Triassic Period, an arid landscape with low relief extended across much of the southwestern part of the North American continent. Some of the most colorful rocks in Colorado are red beds deposited in this environment. These red sediments also became the iconic red rocks of the Colorado Plateau, well displayed at Arches, Canyonlands, and Capitol Reef National Parks and Colorado and Dinosaur National Monuments. These units were deposited in river channels, floodplains, lakes, playas, and mudflats, and then

5

exposure to the atmosphere oxidized iron in the sediments, turning them red. In Colorado, these red bed units include the Chinle, Lykins, and Moenkopi Formations.

The Jurassic Period heralded a change in climate from mudflats to extensive arid dune fields, and then back to a moister climate. The dunes are captured in the Entrada and Navajo Sandstones. The famous dinosaur-bearing Morrison Formation, a geologic unit that spans much of western North America, was deposited in Jurassic time. The Morrison Formation is a red, gray, and green claystone, each color recording environmental conditions during the time of deposition. Greens source from pyrite (iron sulfide) and iron silicate cements bound in reducing conditions (like a swamp or bog), the reds record oxidation of iron-cemented sediments, and the gray and clay-based units were deposited in a poorly drained, swampy, low-energy setting. Dinosaurs lived in these environments, and the accumulating sediments preserved dinosaur remains.

In Cretaceous time, the Western Interior Seaway flooded the continent's interior, including much of Colorado. Sediment accumulated at the bottom of the sea, including about

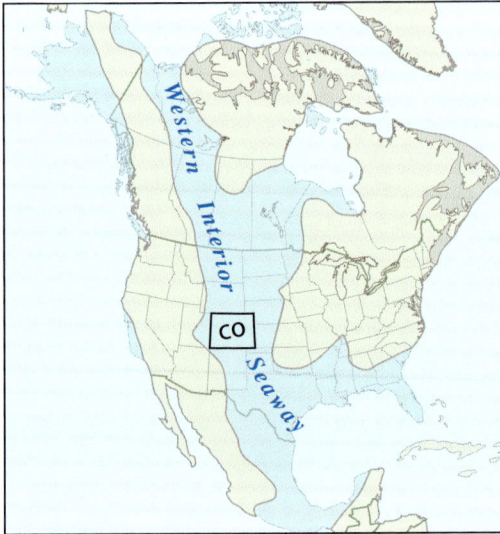

In Cretaceous time, an inland sea stretched north-south across North America, completely inundating Colorado.

700 feet of what is now the Pierre Shale (notable for gas production in the Denver Basin) and the Mancos Shale, named for the western Colorado town of Mancos near its type section. Sea level rose and fell for millions of years until the sea drained completely as tectonism resumed about 75 million years ago. Colorado has not been under seawater since.

The Laramide orogeny began folding the crust about 75 million years ago. Pressures from the collision of tectonic plates along the western edge of North America transferred compression across the continent, buckling the North American plate. The mountain ranges that formed from the compression are long, linear, abrupt warpings of the crust. To think about how this happens, imagine a table with a tablecloth upon it. If you push gently on one end of the fabric, the fabric will move without crumpling; if you keep pushing, eventually the cloth will form ridges and folds, piling up on itself in the middle of the table. The Rocky Mountains in Colorado are the crumpled section of the tablecloth.

During and following the Laramide orogeny, pockets of hot magma bubbled up within the crust over tens of millions of years and slowly cooled to form crystalline rock such as granite. The crust was fractured and heated, and fluids circulating through the rocks nearby were superheated and often became supersaturated with ore minerals. As these supersaturated fluids migrated through the fractured crust, away from the plutons, the waters cooled and ore minerals precipitated out into veins of precious minerals or metal ores, forming the Colorado Mineral Belt. This zone of rich bodies of precious minerals and ores, a swath of land nearly 300 miles in length and about 25 miles across, stretches diagonally from southwest to northeast Colorado, right through the heart of the Rocky Mountains. The most commonly found ores are mixed metal sulfide veins that include the minerals pyrite, galena, sphalerite, and chalcopyrite, from which gold, silver, and copper are derived. The origins of the Colorado Mineral Belt have been long debated. The primary orientation of the band of productive mineral deposits is in line with the northeast-trending pattern in which Proterozoic rocks were accreted to the continent. In addition to the Laramide orogeny, Oligocene volcanism and Rio Grande rifting contributed substantially to the mineralization.

While silver and gold mining are the glittery side of ore extraction in Colorado, uranium and vanadium are equally valuable and useful. In addition to using uranium's radioactivity to generate electricity, uranium has many other uses. Its very long half-life (4,000,000,000 years!) makes it perfect for the radiometric dating of some of the oldest rocks on the planet. Uranium mining in Colorado began in 1872 at Central City. The majority of uranium from the Front Range came from the Schwartzwalder Mine, which produced more than 17 million tons of uranium oxide. In the Uravan mineral belt, an arc-shaped zone in southwestern Colorado and southeastern Utah, mining began in 1898 from a deposit at Roc Creek, Colorado. The Uravan mineral belt was responsible for nearly half the world's uranium mines in the early 1900s, after which major deposits were found in the Democratic Republic of Congo. Mining also began in the 1950s in the Cochetopa mining district near Gunnison and at Tallahassee Creek in Fremont County in northern Colorado, and in the Denver Basin. Uranium mining in Colorado has started and stopped many times as the global price of uranium fluctuated over the past 150 years.

As the Laramide folded and buckled the crust, both high peaks and basins were created. Basement rock along with the sedimentary rock that covered it was elevated. Any high point on the globe is subject to accelerated erosion, and these new high ranges were no exception. The new Laramide highlands were attacked by wind, snow, and rain, which removed much of the sedimentary cover overlying the Proterozoic cores. These eroded sediments washed into and filled ever-deepening basins, including the Denver Basin to the east, the Piceance Basin to the west, the Raton Basin of southeast Colorado, and the San Juan Basin of southwest Colorado and northwest New Mexico. These former basins, now completely filled with sedimentary rock layers, are some of the most economically important geologic features in the West because they host vast reserves of petroleum, natural gas, and oil shales. The majority of the petroleum-bearing rock units in Colorado were deposited in the Western Interior Seaway during Cretaceous time. The remains of flourishing plants and animals sank to the bottom of the sea into stagnating conditions lacking oxygen. As layer after layer of biological material as well as sand and mud were deposited on the seafloor, the temperatures of the buried layers rose, first causing the biological material to decay, then as heat grew, formed waxy kerogens, and eventually, at depths of at least a half mile and over millions of years, to form the hydrocarbons we utilize today.

Natural gas is found in shale and sandstone that were deposited in a coastal plain. Oil and natural gas occur in the Mancos Shale and Mesaverde Group, and oil shale occurs in the Eocene Green River Formation of the Piceance Basin. Most of the petroleum production in the Denver Basin comes from the Cretaceous-age Pierre Shale, with additional production from Permian sandstones. Coal in the Denver Basin is found in the Laramie Formation, also of Cretaceous age. Coal in western Colorado is found mostly in the Mesaverde Group.

While the Laramide orogeny was continuing with its uplift, a calamitous event dealt a blow to life on Earth. A major asteroid impact near the Yucatán Peninsula of Mexico

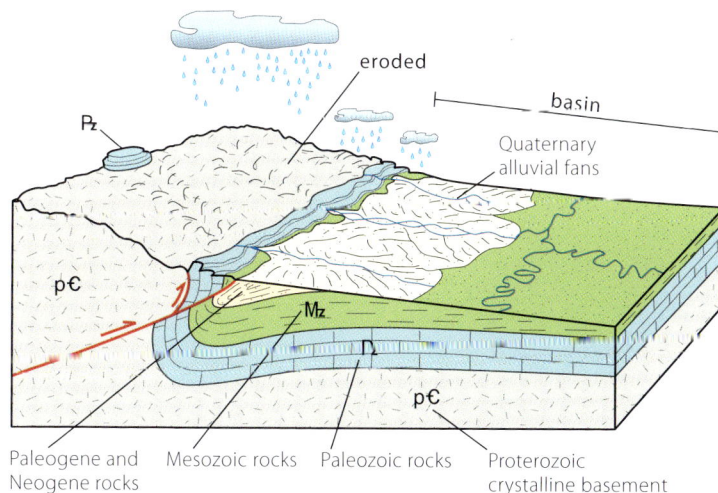

During the Laramide orogeny, old rocks from the basement were thrust to the east along low-angle faults. Sedimentary layers were eroded from the mountains.

caused the sudden extinction of many life forms, including the dinosaurs. Geologists use changes in fossils to determine the boundaries of time periods of the geologic time scale, so the extinction defines the major boundary between Mesozoic and Cenozoic time. The boundary, marked by unusual chemistry and now known as the K-Pg (geologic shorthand for Cretaceous-Paleogene) boundary, is preserved in basin sediments in Colorado and can be viewed at Trinidad Lake State Park. The thin layer was known as the K-T boundary until the Tertiary Period (T) was replaced with the Paleogene and Neogene Periods.

Cenozoic Time

The Laramide orogeny continued into Cenozoic time. The folding, uplift, and magmatism came to an end about 40 million years ago, but the quiescence would not last long. Volcanism began soon after with lava flows from intermediate-composition volcanoes. One of the notable remnants of volcanic eruptions from this period is the Thirtynine Mile volcanics that erupted from volcanoes in the Sawatch Range between 38 and 29 million years ago. One of the volcanic units from this field is the 36.7-million-year-old Wall Mountain Tuff, fragments of which are found at Castle Rock on the east side of the Front Range. The tuff cooled from a pyroclastic flow from a caldera located near Mt. Princeton.

This early phase of volcanism was followed in Oligocene time by the catastrophic eruptions of more than twenty calderas that now define the San Juan volcanic field. Ash flows that erupted from these calderas covered the entire San Juan region with sheets of volcanic tuff between approximately 30 and 25 million years ago. The largest of these calderas, La Garita, was one of the most explosive eruptions in world history, ejecting more than 1,200 cubic miles of volcanic material about 27.8 million years ago. These calderas often erupted more than once: La Garita continued to have at least seven eruptions following its major outpouring, the last about 26 million years ago. Thick layers of volcanic rock including tuff and lava blanketed the landscape, burying older sedimentary rocks and basement rocks. Magmatic fluids circulated through fractures, precipitating minerals and ore.

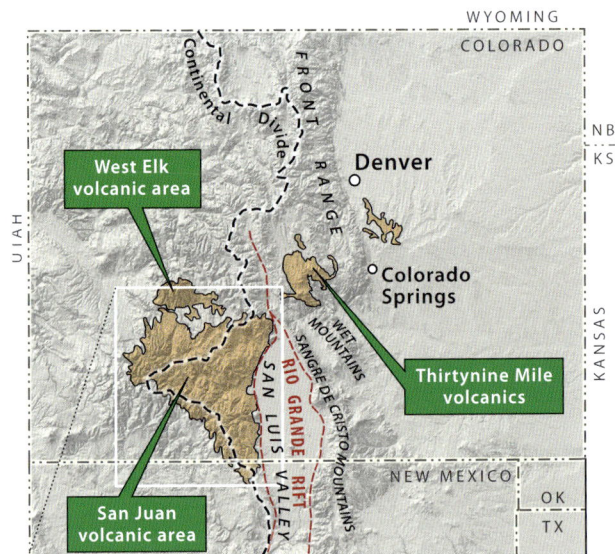

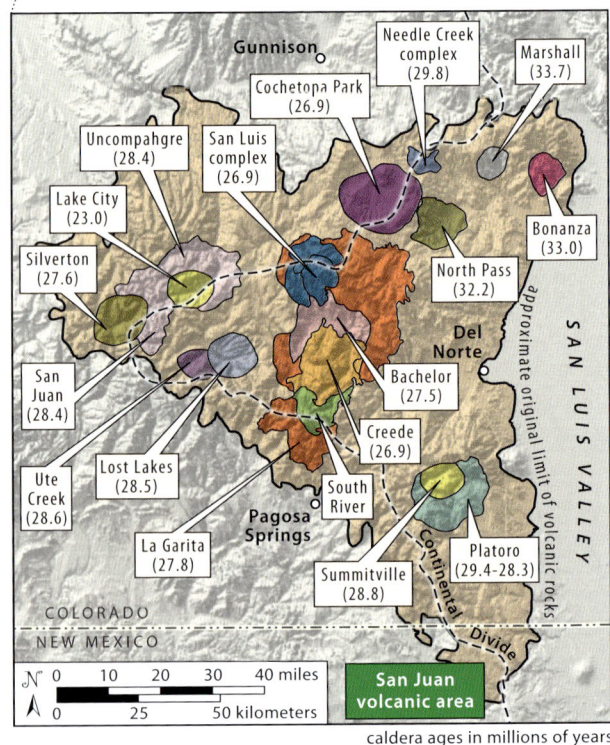

Some Cenozoic volcanic fields in Colorado, with a close-up of the calderas in the San Juan volcanic field. Ages in millions of years.
—Modified from Lipman, 2007; Steven and Lipman, 1976

Beginning about 30 million years ago, tectonic forces attempted to split the North American continent in two, leaving a connected series of rift basins that extend from the Gore Range in northern Colorado south through New Mexico to Texas. Collectively known as the Rio Grande rift, the basins dropped down along faults as the crust thinned and stretched. Down-dropped basins include the upper Arkansas Valley and the San Luis Valley in Colorado.

Major uplift occurred between 30 and 10 million years ago, but the shaping of the topography was not anywhere near done. Erosion attacks every high point on the globe, and the Rocky Mountains and volcanic edifices were no exception. Sediment was deposited both to the west of the Rockies in the rift basins and to the east onto the Great Plains. Erosion was so intense during this time along the Front Range that the basement rock was beveled to near level, and rivers draining the highlands spread huge quantities of sand, gravel, and mud across the Great Plains. So much material was removed from the highlands that isostatic rebound further buoyed up the Earth's crust in the area of removed material.

The Pleistocene Epoch, an extended period of glaciation, began 2.6 million years ago. Ice, usually concentrated at the Earth's poles, expanded into midlatitudes on numerous occasions. Colorado was far enough south to avoid the massive continental ice sheets that spread south from Canada, but it was cold enough that glaciers formed in Colorado's mountains and extended out into valleys. Deposits from only the last two ice ages are present in Colorado; the evidence of earlier ice was removed by glaciers of the last two ice ages. The alpine glaciers in Colorado carved cirques (basins surrounded on three sides by steep rock walls), horns (pointed peaks), and arêtes (sharp ridges) into the hard rock and left large deposits of sediment called moraines.

Numerous hot springs dot the mountainsides and river valleys of Colorado. High heat flow results from ongoing tectonic activity and shallow mantle sources. In the Rio Grande rift, the stretched crust is thin, and hot mantle is close to the surface, heating groundwater. As rain and snowmelt percolate into the crust, they mix with groundwater, and hot rocks at depth heat the water. This heated water rises to the surface along cracks and faults, bringing with it minerals dissolved from bedrock. The rotten-egg scent of many hot springs is from hydrogen sulfide gas, generated by the presence of anaerobic bacterial conversion of sulfur to sulfide.

A glacial lake, or tarn, rests in a glacial cirque near the top of Pagoda Mountain in Rocky Mountain National Park. The U-shaped valley in the distance below the cirque is the telltale sign of a former glacier. —Marli B. Miller photo

FRONT RANGE AND GREAT PLAINS

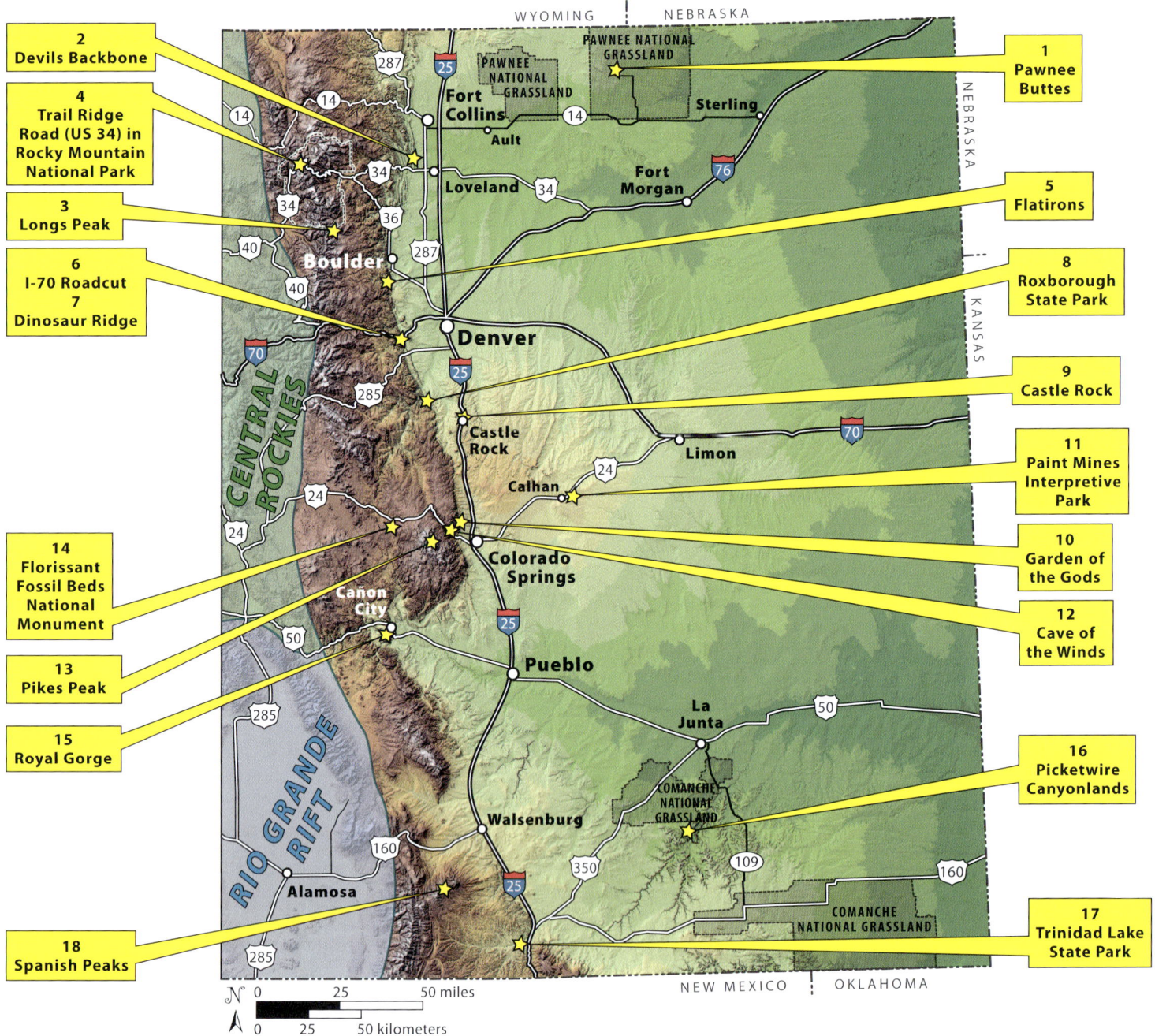

WYOMING | NEBRASKA

2
Devils Backbone

4
Trail Ridge Road (US 34) in Rocky Mountain National Park

3
Longs Peak

6
I-70 Roadcut
7
Dinosaur Ridge

14
Florissant Fossil Beds National Monument

13
Pikes Peak

15
Royal Gorge

18
Spanish Peaks

1
Pawnee Buttes

5
Flatirons

8
Roxborough State Park

9
Castle Rock

11
Paint Mines Interpretive Park

10
Garden of the Gods

12
Cave of the Winds

16
Picketwire Canyonlands

17
Trinidad Lake State Park

PAWNEE NATIONAL GRASSLAND

PAWNEE NATIONAL GRASSLAND

Sterling

Fort Collins

Ault

Loveland

Fort Morgan

Boulder

Denver

Limon

Castle Rock

Calhan

Colorado Springs

Cañon City

Pueblo

La Junta

Walsenburg

COMANCHE NATIONAL GRASSLAND

COMANCHE NATIONAL GRASSLAND

Alamosa

CENTRAL ROCKIES

RIO GRANDE RIFT

NEBRASKA

KANSAS

NEW MEXICO | OKLAHOMA

N

0   25   50 miles

0   25   50 kilometers

The Front Range extends south into northern New Mexico and north into southern Wyoming. The striking rise from the grassy plains to snow-capped peaks over 14,000 feet in elevation is awe inspiring in any season, spurring questions of how this dramatic landscape could have been created. Two great mountain building events crumpled the Earth's crust into mountains here. The first, the Ancestral Rockies orogeny, occurred 320 to 290 million years ago, and those mountains eroded long ago and were covered with layers of sedimentary rock. During the second, the Laramide orogeny, blocks of rock were lifted upward into large folds. Erosion removed the softer, overlying sedimentary rock, exposing granite and gneiss. Steeply tilted fins along the eastern edge of the Front Range and scattered remnants within the high mountains are all that remain of the once continuous sedimentary cover.

The tilted rock layers dip into the Denver Basin, a large asymmetrical syncline, or downwarping, on the eastern side of the Front Range. The basin's deepest part, reaching 13,000 feet in depth, is along its western edge, very near the Front Range, and it shallows out gradually to the east. The sedimentary rock that fills the basin eroded from the Ancestral Rocky and Laramide uplifts over millions of years and accumulated in many different depositional environments, thus preserving a reverse history of the repeated rise of the Rockies. The eastern plains of Colorado slope very gradually away from the Front Range.

The Denver Basin reaches from Pueblo north into southern Wyoming, so Denver and other Front Range cities sit atop it. Oil and gas production began in the Denver Basin in 1901, when the Pierre Shale was tapped for oil at the historic McKenzie well near Boulder. Natural gas is produced from the Pierre Shale and Niobrara Formation, both of Cretaceous age. Small amounts of coal have also been mined, though not to the economic extent of oil and natural gas.

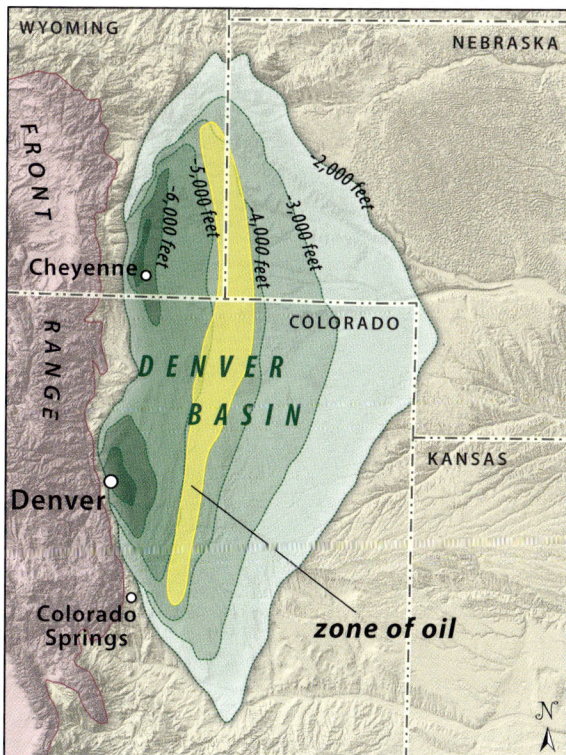

Oil is produced from Cretaceous rocks in the Denver Basin.

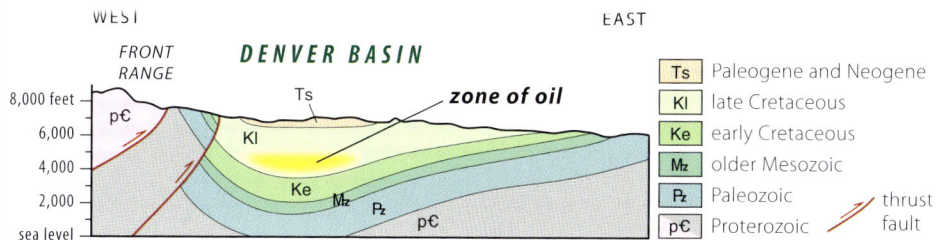

1 Pawnee Buttes
Looking into the High Plains

Pawnee Buttes, erosional remnants of the High Plains, rise 300 feet above the surrounding plains in the Pawnee National Grassland about 70 miles east of Fort Collins. Following the Laramide orogeny and uplift of the Rockies, erosion removed sedimentary rock from the highlands and deposited the sediments in a gently sloping plain to the east of the mountain front, forming the High Plains. This broad surface that extends from New Mexico to Canada has been warped, tilted, and dissected by rivers since its formation. The Pawnee Buttes preserve a more complete depositional history than the surrounding piedmont, which has been cleared of Cenozoic sediments. The buttes expose, from bottom to top, the White River Group, the Arikaree Formation, and the Ogallala Formation. The latter unit of sandstone and conglomerate lies at the very top of the eastern butte.

The Ogallala Formation was deposited in large alluvial fans as water and sediment were shed off the eastern flank of the rising Rocky Mountains. The Ogallala Formation and underlying Arikaree Formation consist of poorly cemented and unconsolidated sandstone, siltstone, shale, clay, and gravel. Water can flow through and within these units because there are no confining barriers and little cement. Farther east, where these sediments lie beneath the plains, they form the Ogallala Aquifer. Also known as the High Plains Aquifer, groundwater in the Ogallala Formation and underlying Arikaree Group sandstones has made life as we know it possible in the plains, supplying vast amounts of water to the people and industries there. Just over one quarter of the irrigated agricultural land in the United States relies upon water from the Ogallala Aquifer.

The Oligocene through Miocene formations exposed at Pawnee Buttes preserve mammalian fossils that have been studied in detail. The Brule Formation of the White River Group contains fossils of at least ninety mammal species, including rhinoceros, fanged giant pigs (oreodonts), horses, camels, small deer, and saber-toothed cats. There are also fossils of fish, amphibians, reptiles (alligators, turtles) and birds, as well as gastropods. The Arikaree Formation also hosts many agates, though collecting fossils and rocks from the Pawnee National Grassland is prohibited. Pawnee Buttes are in the eastern unit of the grassland, a remote region with primarily gravel and dirt roads.

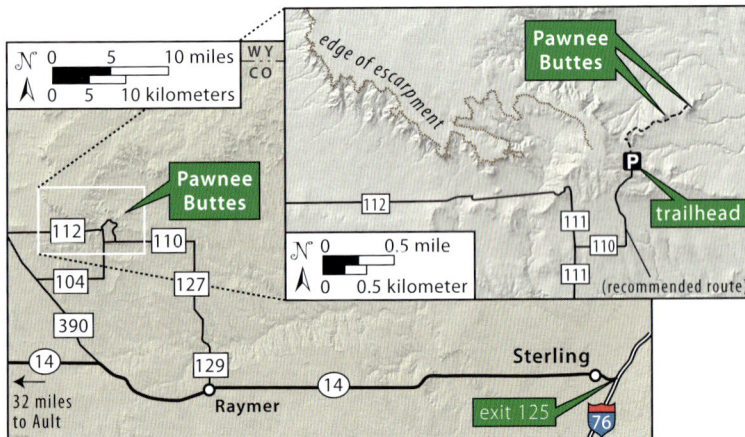

The National Forest Service manages the Pawnee National Grassland from an office in Ault on CO 14. Check their website for information, directions, and current information.

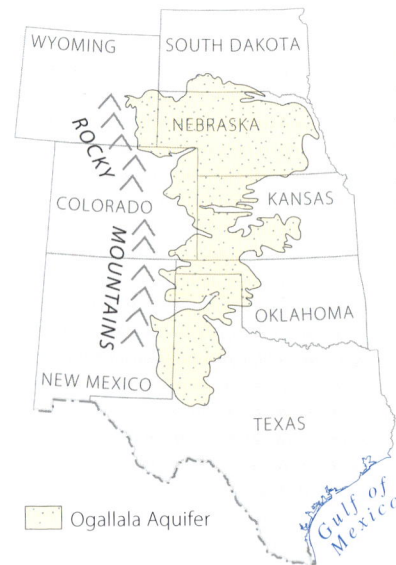

Groundwater in the Ogallala Aquifer is contained within sediments shed from the Rocky Mountains to the west.

Ogallala Aquifer

Close-up view of some poorly cemented conglomerate of the Arikaree Formation at Pawnee Buttes. Pocket knife for scale in upper left.
—Marli B. Miller photo

A person stands on the Arikaree Formation, which forms a resistant cap over softer rock of the White River Group. In the background, the White River Group forms the sloping and vertical sides of the Pawnee Buttes. Arikaree Formation caps the buttes, with a veneer of conglomerate of the Ogallala Formation barely present on the top of the eastern butte (right).
—Marli B. Miller photo

13

The high, narrow ridge of Devils Backbone, a resistant unit of Dakota Group sandstone, tilts slightly westward above the valley floor. View to the north. —Magdalena S. Donahue photo

2 Devils Backbone Open Space
Spine of the Dakota Hogback

Devils Backbone is a nearly 2-mile-long erosional remnant of the Dakota Hogback, a ridge of Dakota Group sandstone that extends along the very eastern edge of the Front Range from southern Colorado to southern Wyoming. Some sections of the ridge are more dramatic than others, and west of Loveland the sandstone rises about 200 feet above the surrounding landscape like a curving spine with knobby vertebrae. Here, a resistant, tan-colored sandstone unit is tilted to nearly vertical, and rock layers on either side of it have eroded away, leaving behind only a rugged, 15-foot-wide layer of coarse-grained sandstone. Gaps and arches such as the iconic Keyhole arch form where water, snow, and ice combine to crack and remove rock material. You can explore Devils Backbone via many hiking trails in the Devils Backbone Open Space, protected by Larimer County Parks and Open Lands. Look for the Keyhole along the Wild Loop Trail.

The Dakota Group that makes up the spine of the backbone are ancient beach sands and other sediments, deposited in river deltas and along the shore of the Western Interior Seaway in the early Cretaceous Period. Soon after deposition, the sea expanded and inundated the region, and marine deposits such as shale became more common. Compression during the Laramide orogeny caused the originally flat-laying sedimentary rock to be warped upward as the Rockies grew. Erosion has since removed much of the softer, more erodible rock units like the shales, leaving the strong, resistant sandstone to stand high in the landscape.

A nearby gypsum mine and plaster mill operated from the 1887 to 1965 and are visible from trails on the east side of the backbone. The gypsum, precipitated by evaporation from a shallow sea during the Triassic Period, occurs within the Lykins Formation.

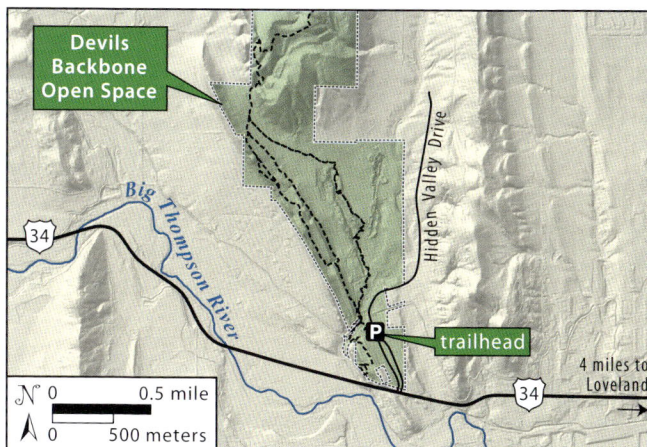

Devils Backbone Open Space lies 4 miles west of Loveland along US 34.

The Keyhole arch in the Devils Backbone. As the near-vertical prominence of stone weathers and erodes over time, larger pebbles fall from the main rock body, creating holes in which water and snow can collect. Freeze-and-thaw action on these small holes over many years enlarges them until they collectively penetrate the rock, becoming an arch. —Magdalena S. Donahue photo

15

3 Longs Peak
The Crown of Rocky Mountain National Park

Rocky Mountain National Park bookends the geologic history in Colorado: glaciers of the recent ice age that ended just 15,000 years ago sculpted and carved billion-year-old intrusive and metamorphic rocks. The towering peaks are some of the oldest crystalline basement rocks in Colorado, including the 1.4-billion and 1.7-billion-year-old granites that intruded even older gneiss and schist. Similar-aged crystalline rocks form the heart of the central Rocky Mountains as they stretch from New Mexico north through Colorado and Wyoming. The granites and other crystalline rocks at Rocky Mountain National Park were uplifted during the Laramide orogeny, with more uplift occurring as material was removed by erosion.

The most-hiked summit in the park is 14,259-foot Longs Peak, named after early explorer Major Stephen Long, who explored the peak in 1820. Although very popular, this 7.5-mile one-way trail is difficult. The final push to the summit is very exposed and not for the faint of heart. Longs Peak is largely formed of the Silver Plume Granite, which intruded the 1.7-billion-year-old Longs Peak–St. Vrain batholith about 1.4 billion years ago.

The popular Chasm Lake, just over 4 miles from the trailhead (8.5 miles in total round-trip length), fills a glacial cirque, a hollow carved out of a mountainside by the cutting power of a glacier. Cirques open on one side where the glacier flowed downhill, and many hold lakes in their basins. Most of the rock visible at Chasm Lake is the light-gray Silver Plume Granite. Metamorphic biotite schist and gneiss, remnants of the original country rock, form the inclined dark layers of the Ships Prow, a prominent feature on the south side of the lake.

Longs Peak, the main peak in the center of this image above the light-gray cliff face, stands above the cirque that hosts Chasm Lake. The inclined layers in the prominence on the left side of the photo, called the Ships Prow, are layers of lighter granite and darker, older metamorphic rock. —Magdalena S. Donahue photo

A cairn, or pile of stones, marks the trail toward Longs Peak. At and above treeline, trails can be hard to see and follow. Human traffic can damage fragile alpine ecosystems, so look for cairns to stay on trails. This cairn is made up of the 1.4-billion-year-old Silver Plume Granite; note the prominent white-pink potassium feldspars and gray quartz minerals.
—Magdalena S. Donahue photo

Chasm Lake hidden in cirque

Lateral moraines, visible as tree-lined ridges on either side of the canyon at bottom left, were deposited along the sides of the glacier that once began at the base of Longs Peak (high point at right). Metamorphic rock in the foreground.
—Marli B. Miller photo

moraine

moraine

4 Trail Ridge Road
Alpine Glaciation, Past and Present

Trail Ridge Road provides an opportunity to those of us who want to experience wildflowers blooming in the alpine tundra but don't have the strength to hike to the top of high-elevation mountains. Open only in the summer, Trail Ridge Road (US 34) travels more than 48 miles through Rocky Mountain National Park from Estes Park to Grand Lake, reaching a high point at 12,183 feet. The route, the highest through-going paved road in Colorado, rises more than 4,000 feet in elevation from either side of the pass, and drivers travel through several ecosystems ranging from aspen and ponderosa pine, to subalpine fir and spruce, eventually reaching alpine tundra that stretches above tree line. The Colorado state flower, the Rocky Mountain columbine, can be seen blooming abundantly throughout this landscape, even into the boulder-strewn alpine tundra. The road crosses the Continental Divide at 10,758-foot Milner Pass.

The peaks we see along Trail Ridge Road are composed mostly of Proterozoic igneous and metamorphic rock. Westward from Iceberg Pass, however, you can see younger volcanic rock at Iceberg Lake, Specimen Mountain, and the Never Summer Mountains. The striking peaks (horns), knife-edged ridges (arêtes), and scooped-out lake-filled bowls (cirques) are all the result of ice eating away at the rock. The area that is now Rocky Mountain National Park was largely covered in snow and ice during the major glaciations of the Pleistocene Epoch. Evidence of the last two ice ages, the Bull Lake ice age from 130,000 to 95,000 years ago and the Pinedale at approximately 23,500 to 15,000 years ago, exist in Colorado's mountains. During glaciations, accumulating snow at high elevations becomes increasingly thick, compacting and turning into ice. Year after year, these ice patches become heavier and are pulled by gravity and pushed downhill by their own growing weight. Rocks embedded in this hard plug of ice carve and scratch deep gouges into the rock over which the glacial ice slides. The debris that is ground up and carried by a moving glacier gets deposited in moraines, huge piles of rubble that form along the sides and around the toe of the glacier where the ice is actively melting.

Today, a few glaciers still occupy cirques in the park, and well-established hikes visit Andrews Glacier and Tyndall Glacier. Several snowfields look like glaciers from a distance, but these are simply areas of year-round snow cover and lack the flowing movement and ice-forming nature of a true glacier. Ice is still at work today where water freezes and then expands, heaving upward and prying rocks apart. The paved road is particularly susceptible to frost heaving.

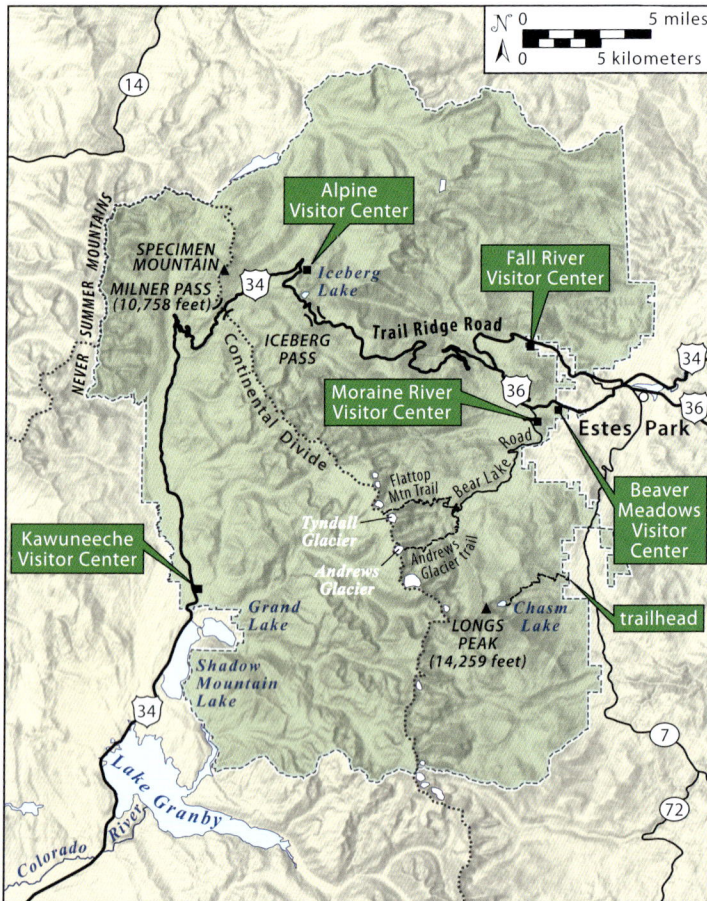

Rocky Mountain National Park, one of the largest national parks in the country, lies 30 miles west of Loveland and 35 miles northwest of Boulder.

Andrews Glacier rises above the cirque basin into which it once flowed. Andrews Glacier can be hiked to via a trail, just over 4 miles one way, that begins at the end of Bear Lake Road. —Marli B. Miller photo

Patterned ground, formed when soil and loose rock freeze and then heave upward, can be seen in the alpine tundra along Trail Ridge Road near Iceberg Lake. —Marli B. Miller photo

19

The iconic Flatirons, viewed in morning light, stand above the plains near Boulder. —Magdalena S. Donahue photo

5 Flatirons at Boulder
Resistant Fins of Fountain Formation

The iconic Flatirons at Boulder are recognizable from miles away and have become an image synonymous with the Colorado Front Range. These towering, triangular rock features, named First Flatiron, Second Flatiron, and so on from north to south, are composed of the Fountain Formation, a pink sandstone with large particles of eroded granite and grains of quartz and feldspar. These large grains provide traction to the multitude of rock climbers who ascend the faces when falcons are not nesting.

Rivers flowing from the Ancestral Rocky Mountains deposited this coarse-grained sediment during the Pennsylvanian and early Permian Periods. Feldspar grains break down quickly to clay, so the fact that they were still intact when deposited indicates they were not carried far from the mountains. They likely came to rest in alluvial fans and braided streams. The reddish-pink color of the rock is due both to the pink feldspars from the eroded granites and gneisses of the Rocky Mountain basement rock and to the purple-red hematite (iron oxide) cement that binds the unit together.

The Fountain Formation was deposited directly on a surface eroded onto the 1.7-billion-year-old Idaho Springs Formation and 1.7-billion-year-old Boulder Creek Granite, some of the basement rocks that form the core of the Rocky Mountains. The contact between the Proterozoic crystalline basement rock and the overlying Paleozoic sedimentary rock is known as the Great Unconformity, due to the huge amount of the geologic record that is missing. More than

1 billion years of geologic history—of deposition, uplift, and erosion—is not known from here because no rocks are preserved. The Great Unconformity is present in many places throughout the West, including the Grand Canyon.

The Fountain Formation and the overlying younger sedimentary units were tilted dramatically during the Laramide orogeny (75 to 40 million years ago), as compression folded and buckled the overlying sedimentary rock layers above the Proterozoic core of the burgeoning Rocky Mountains. Highly resistant to erosion, fins of the Fountain Formation stand tall, while overlying units to their east have eroded away, all sloping east into the Denver Basin.

The Fountain Formation, coarse grains deposited in ancient alluvial fans, contains cross beds that formed as channels shifted on the fan surface. —Marli B. Miller photo

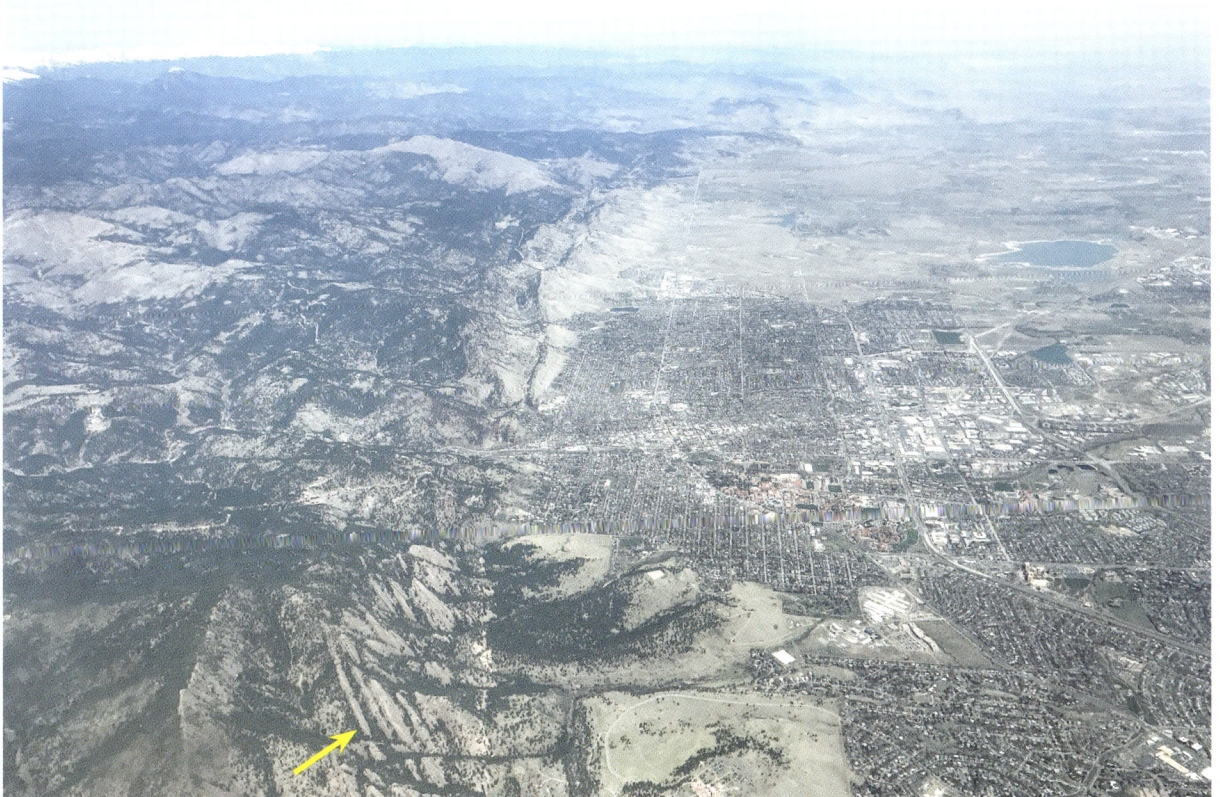

The city of Boulder sits at the base of the Flatirons, where the plains meet the Front Range. Arrow points to the location of the Great Unconformity behind the Flatirons. —Marli B. Miller photo

21

View looking north at the Dakota Hogback, known as Dinosaur Ridge here, as it curves along the Front Range. I-70 cuts across the bottom of the photo. —Marli B. Miller photo

6 I-70 Roadcut
Dakota Hogback Dips into Denver Basin

Colorado's highways have numerous roadcuts, but only one is its own geologic site, complete with parking lots and interpretive trails. The steeply tilted layers exposed in the geologically famous I-70 Roadcut are part of a hogback ridge, a long, linear feature with steep sides and a narrow crest. Hogbacks are so named because they resemble the back of a hog. They stand above the nearby landscape due to differences in erodibility: the more resistant rock layers at the core of the hogback remain in place while softer layers wash away. Tilted rocks usually form a hogback with a gentler side up the slope of the bedding and a very steep opposite side. Devils Backbone (site 2) is another section of this classic hogback.

The Dakota Hogback is named for the Cretaceous-age Dakota Group, a sandstone formation that was deposited about 120 million years ago. Its buff and black hues highlight the eastern end of the hogback (South Platte and Lytle Formations). The rocks get older to the west as you continue westward toward the mountains. The steeply dipping but less visible rocks in the valley on the west side of the hogback are the 150-million-year-old Morrison Formation, 160-million-year-old Ralston Creek Formation, 200-million-year-old Lykins Formation, and the 250-million-year-old Lyons Formation.

If you take exit 259 and park at the park-n-ride lots on the west side of either the north or south roadcut, you can walk along a designated trail and examine these rocks personally. The northern roadcut also hosts informative interpretive signs. See map for site 7.

Black shale of the Cretaceous Dakota Group with interpretive sign along the northern walkway. —Marli B. Miller

Striking colors pop out in the Dakota Hogback just west of Denver. Beds of bright red and yellow sandstone, red and white siltstone, and black shale are part of the Dakota Group. In the far western end (right) of the hogback, the red Morrison Formation becomes visible. The engineered terraces prevent rockfall from reaching the highway. —Magdalena S. Donahue photo

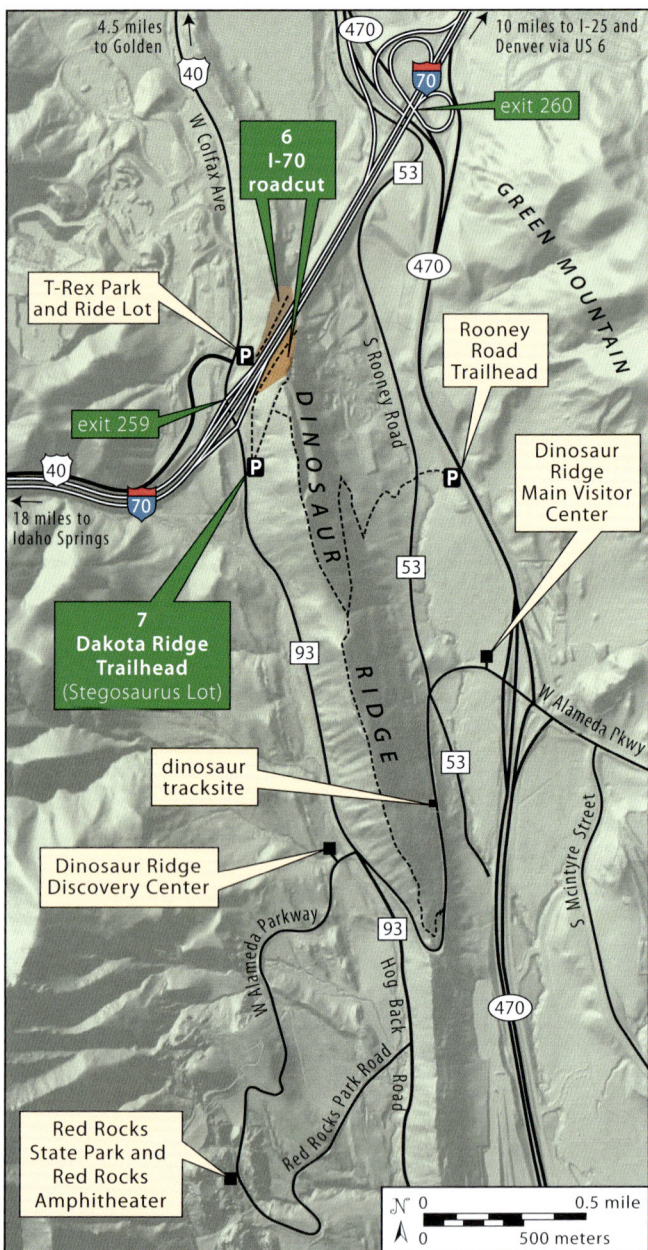

The I-70 Roadcut can be accessed by taking exit 259 to Morrison from I-70. For westbound travelers, an immediate right turn into the T-Rex Park-and-Ride parking lot provides excellent access to the walking trail on the north side of the roadcut. To access the south side of the roadcut, take the same exit, cross over I-70 and park in the Stegosaurus Lot, from which the roadcut and the Dakota Ridge Trail (urban trail) can be used to access Dinosaur Ridge.

Dinosaur Ridge
Fossils and Tracks below the Dakota Hogback

In the 1870s as miners scoured the mountains for gold and silver and a new railroad brought commerce to Denver, other people were interested in much different discoveries. Fossils, including those of *Apatosaurus*, *Diplodocus*, *Allosaurus*, and *Stegosaurus* (the Colorado State Fossil), were discovered in the Morrison Formation in 1877 by Arthur Lakes. Digs for the fossils were conducted at what was referred to as Morrison Quarry No. 5 at what is now known as Dinosaur Ridge, near Red Rocks Amphitheater in Morrison. This world-famous dinosaur fossil locality has been designated a National Natural Landmark.

The ridge, a hogback formed by differential erosion of tilted sedimentary rocks, exposes sedimentary layers from Late Jurassic to Cretaceous age. Rocks of the resistant Cretaceous-age Dakota Group form the ridge crest, while more highly erodible sands, muds, and silts of the underlying Jurassic-age Morrison and Ralston Creek Formations lie in the lower western slope of the ridge. The Dakota Hogback extends the length of the Front Range, forming an abrupt geologic breaking point between the Great Plains to the east and the Front Range to the west.

Dinosaur bones are found in abundance in the Jurassic rocks; it appears that dinosaurs were washed into a braided stream channel during a flood event about 150 million years ago. Quickly covered by sediment, the dead animals became preserved as fossils. Soft tissues decomposed, and dissolved minerals crystalized in the bones, turning them to stone.

Dinosaur Ridge also includes a dinosaur tracksite discovered in 1937 during highway construction. More than 300 dinosaur tracks, including those from iguanodons, occur in early Cretaceous Dakota Group sandstones. These dinosaurs were walking along the coastline sands of the Western Interior Seaway 30 million years after their Jurassic ancestors were buried.

The red rocks of Red Rocks State Park and Red Rocks Amphitheater, farther west, are the Fountain Formation of Pennsylvanian to Permian age. The Great Unconformity is exposed in Red Rocks State Park and noted by a plaque at the edge of the parking lot. Here, the Fountain Formation rests unconformably upon Proterozoic-age gneiss. The Fountain Formation is also visible as the Flatirons in Boulder, and in fins at Roxborough State Park, Garden of the Gods, and Red Rock Canyon Open Space.

The Great Unconformity (at sign) is labeled at the edge of the upper parking lot in Red Rocks Park. The Fountain Formation, the bumpy, gravelly layered unit (left side of photo), was deposited 300 million years ago on an eroded surface of fractured, 1.7-billion-year-old Proterozoic gneiss (right side of photo). —Marli B. Miller photo

Fossilized dinosaur bones at the Dinosaur Ridge quarry. Most of the bones were removed, but a few remain, and some, like this one, were returned to the quarry after removal. —Marli B. Miller photo

Ancient footprints at Dinosaur Ridge. Here, the footprints are filled with dark paint to help visitors see them. —Magdalena S. Donahue photo

Roxborough State Park
Hogbacks and Linear Valleys

Roxborough State Park is probably the best place to see the Paleozoic to Mesozoic sedimentary sequence of Colorado's Front Range. The rocks dip about 60 degrees toward the northeast, so they are all exposed over about 1 mile. They're mostly unfaulted, easily accessible by numerous trails, and have eroded to create a series of parallel hogback ridges and valleys. You can also hike to see the local basement rock, which consists mostly of Proterozoic granite and granitic gneiss.

The three large hogbacks consist of the Pennsylvanian Fountain and Permian Lyons Formations, and the various rocks of the Cretaceous Dakota Group. In between them, erosion of the weaker rocks formed linear valleys, parallel to the horizontal trend of the rock layers. Similar to the Fountain Formation at Red Rocks State Park next to Dinosaur Ridge (site 7) or Garden of the Gods (site 10), the Fountain here consists mostly of red sandstone and conglomerate that was eroded from the rising Ancestral Rockies. The Lyons Formation here likely reflects deposition in coastal areas, including sand dunes, and the Dakota Group includes a wide range of rocks formed in swamps and tidal flats, a harbinger of flooding of the landscape to produce the Western Interior Seaway in Cretaceous time.

Other units, including the Lykins and Morrison Formations, and the Pierre Shale and Fox Hills Sandstone, show up only sporadically through the landscape but tell their own stories. The red Lykins Formation, for example, includes rocks of both Permian and Triassic age and so records an arid, hot landscape before and after the great Permian-Triassic extinction event. The Morrison Formation, which accumulated in a floodplain setting, is famous for its dinosaur fossils. The Pierre Shale was deposited in the Western Interior Seaway, and the Fox Hills Sandstone was deposited later along the seaway's coast as the water retreated.

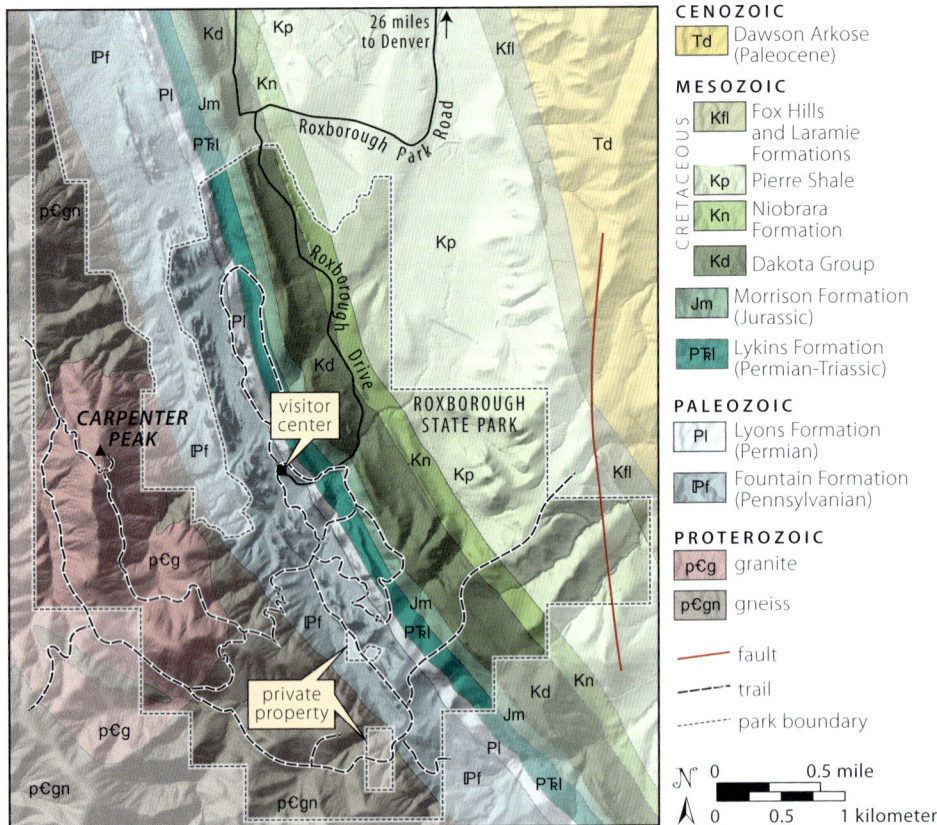

Geologic map and trails of Roxborough State Park southwest of Denver.

Fins of the Pennsylvanian-age Fountain Formation.
—Marli B. Miller photo

View northward down the valley between hogbacks of the Fountain Formation (left) and Lyons Formation (right, foreground). The high ridge on the far left of the photo consists of Proterozoic basement rock. —Marli B. Miller photo

9 Castle Rock
A Story in Conglomerate

Careful study of the pebbles, cobbles, and boulders contained within a conglomerate can tell you a lot about the geologic history of an area. In a striking outcrop in Rock Park above the city of Castle Rock, a conglomerate forms the upper three-quarters of Castle Rock, an erosional remnant. The unit, known as the Castle Rock Conglomerate, contains a wide range of rock types, including granite, quartzite, Fountain Formation conglomerate, and Wall Mountain Tuff. The granite and quartzite were brought from the Front Range mountains, and the 36.7-million-year-old Wall Mountain Tuff vented from the Mt. Princeton caldera approximately 90 miles southwest of Castle Rock in the Sawatch Range, south of the town of Buena Vista. The Castle Rock Conglomerate must be younger than the tuff because it includes pieces of it. Also found within the conglomerate are rare beds of mudstone that contain bones of brontotheres, a large hooved, horned mammal that became extinct 35.5 million years ago. The conglomerate, therefore, must be older than or about the same age as the extinction date.

The Wall Mountain Tuff, which cooled from an ash flow and was welded together by its own heat, is a pink, low-silica rhyolite with abundant sanidine feldspar and plagioclase minerals. The welded tuff is extremely resistant to erosion and caps many of the mesas and buttes in the Castle Rock area. The presence of the tuff as a caprock is an excellent example of inverted topography: when the

Castle Rock, standing above the surrounding city of Castle Rock, is an erosional remnant of resistant Castle Rock Conglomerate. You can see the active breaking down of this high point in the jumble of boulders adorning the slopes. The yellowish Dawson Arkose forms the vegetation-covered hillslopes. —Marli B. Miller photo

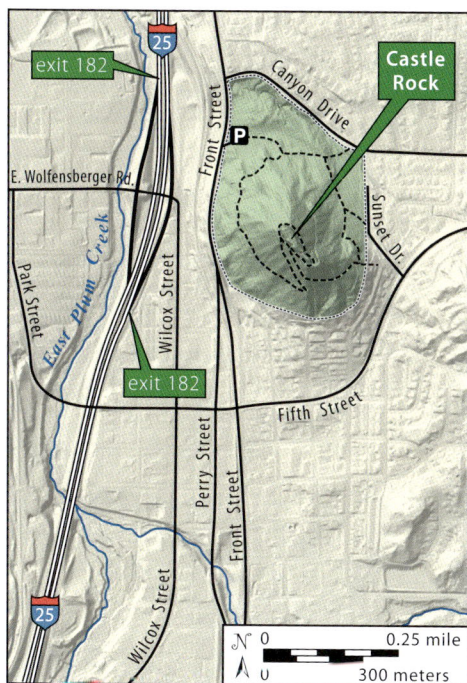

A parking lot on Front Street provides access to trails that circle and climb to the base of Castle Rock.

The Castle Rock Conglomerate (top), a unique unit of jumbled, angular clasts that range in clast grain size from tiny sands to large cobbles, was deposited by fast-flowing streams over the top of the Dawson Arkose (bottom). Note the faint bedding visible at top. —Marli B. Miller photo

The contact of Castle Rock Conglomerate, containing large clasts of Wall Mountain Tuff, with underlying Dawson Arkose (smooth whitish rock next to path). Note how the contact between the two rock units, an erosional unconformity, is irregular. —Marli B. Miller photo

tuff erupted, it flowed down and cooled in the lowest areas of valleys; subsequent erosion removed the surrounding softer sedimentary rocks, leaving the welded tuff high in the landscape, inverting its original depositional geometry. We know the Wall Mountain Tuff was not deposited at Castle Rock because the younger Castle Rock Conglomerate was deposited directly on the much older Dawson Arkose.

The Dawson Arkose, a light-colored sedimentary rock, is exposed at the base of Castle Rock. At the top of the Dawson Arkose is an erosional unconformity, a flat surface eroded before deposition of the overlying Castle Rock Conglomerate. Look for a distinct change in rock types along a tilted horizon about 5 to 15 feet above the base of Castle Rock. Arkose is a type of sandstone containing larger grains of granite and feldspar, not just the quartz sand that is more typical of common sandstones. The granite clasts have been determined to be Pikes Peak Granite, so we know the Dawson was deposited by streams flowing eastward from the uplifting Rocky Mountains. Based on fossils in thin mudstone beds found at other sites, geologists know the Dawson Arkose was deposited during the Laramide uplift, about 60 million years ago.

10 Garden of the Gods
Faulted Fins of Lyons Formation

In contrast to the unfaulted edge of the Front Range at Roxborough State Park (site 8), the Garden of the Gods area of Colorado Springs is highly faulted, with steeply dipping and, in some places, overturned sedimentary rocks. On the northwest side of the park, the Rampart Range fault cuts through outcrops of Pennsylvanian-age Fountain Formation to mark the upthrown edge of the Front Range. To the east, the younger rocks are offset by numerous smaller faults that appear related to this larger structure but moved rocks in the opposite direction, back toward the Front Range. These faults are called back thrusts. As these shuffled rocks erode, they leave behind a seemingly random assortment of near-vertical, multicolored fins in a garden-like setting.

The western two-thirds of the park is composed of the Fountain Formation, sediment shed onto alluvial fans and deltas from the rising Ancestral Rocky Mountains during the Pennsylvanian Period. You can inspect its red and white conglomerates and mica- and feldspar-rich sandstones on

View looking southward along vertical fins of Fountain and Lyons Formations. —Magdalena S. Donahue photo

31

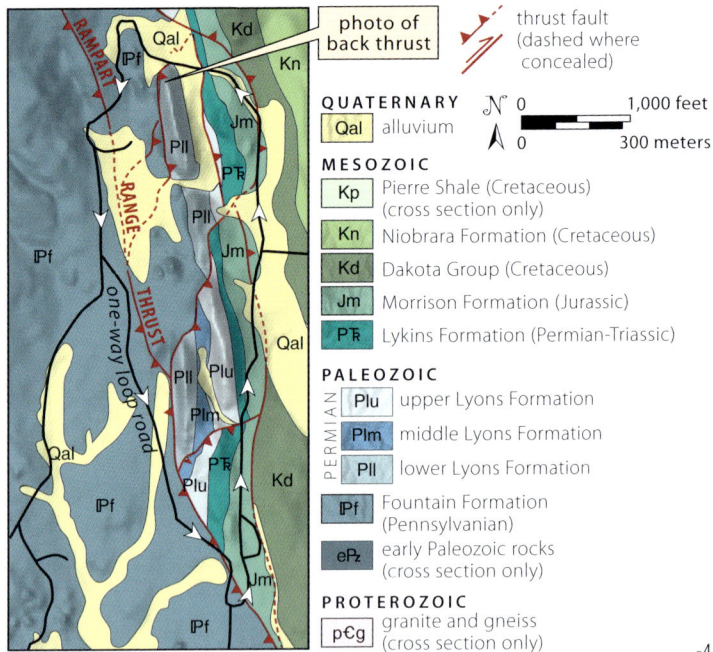

Geologic map of part of the Garden of the Gods, a park on the northwest side of Colorado Springs. —Geologic map compiled from Keller and others, 2005; Morgan and others, 2003; and Siddoway and others, 2013; cross section from Sterne, 2006

Map legend:

photo of back thrust

thrust fault (dashed where concealed)

QUATERNARY
- Qal alluvium

MESOZOIC
- Kp Pierre Shale (Cretaceous) (cross section only)
- Kn Niobrara Formation (Cretaceous)
- Kd Dakota Group (Cretaceous)
- Jm Morrison Formation (Jurassic)
- PŦ Lykins Formation (Permian-Triassic)

PALEOZOIC
- Plu upper Lyons Formation
- Plm middle Lyons Formation
- Pll lower Lyons Formation
- Pf Fountain Formation (Pennsylvanian)
- ePz early Paleozoic rocks (cross section only)

PROTEROZOIC
- pЄg granite and gneiss (cross section only)

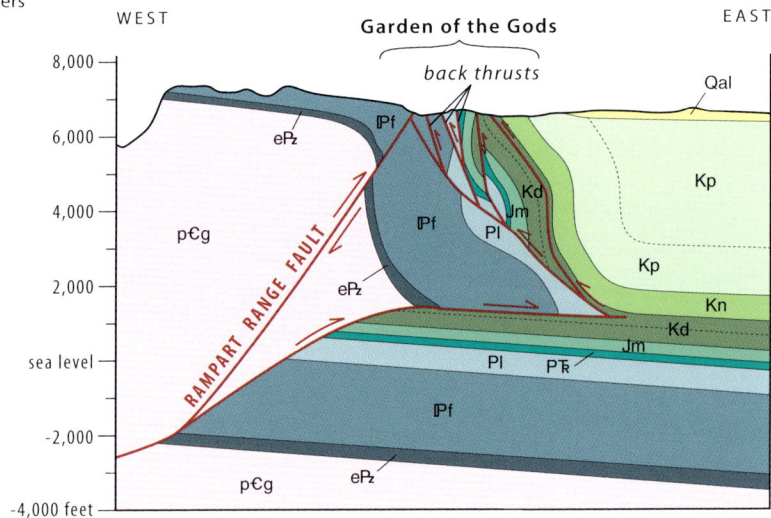

West — Garden of the Gods — back thrusts — East

RAMPART RANGE FAULT

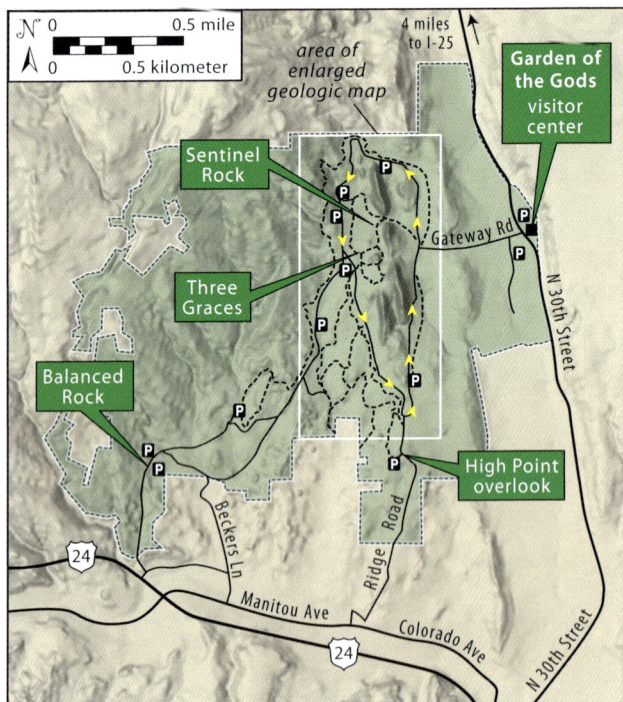

The main loop road (yellow arrows) is a one-way road with occasional parking areas and numerous hiking trails that explore the rock formations.

most of the park's hiking trails. The beds dip shallowly on the west side of the park, where they erode into low ridges and some stand-alone rocks.

The tall fins of the younger Lyons Formation, the park's main attraction, lie on the east side of the park, and some reach more than 100 feet high. The Lyons consists of three parts: a red sandstone at its base, a red conglomerate that resembles the Fountain Formation in its middle part, and a sandstone at its top, which is light gray at the Garden of the Gods. The sandstones formed mostly as sand dunes during the Permian Period and show some beautiful examples of large-scale cross beds. The light-gray sandstone lost its red color to circulating fluids, possibly during the Laramide orogeny. Both sandstones are resistant to erosion and form tall fins. The conglomerate, which was deposited by streams and may mark a period of later uplift of the Ancestral Rockies, erodes more easily to form a valley between the fins. At the north end of North Gateway Rock, you can see the red Lyons Formation faulted against overturned Fountain Formation along one of the park's classic back thrusts.

As you go eastward from the high fins, you encounter progressively younger rocks, forming alternating ridges and valleys depending on their resistance to erosion. Also offset by faults, these younger rocks don't outcrop in a predictable way. The mostly red Lykins Formation likely formed in arid, coastal mudflats during the Permian and Triassic Periods. The Morrison Formation, which contains purple and green siltstone and mudstone and a gray gypsum evaporate with a crusty surface, formed in floodplain and lake environments during the Jurassic Period. The Dakota Group, so beautifully exposed at Devils Backbone (site 2), along I-70 (site 6), and at Dinosaur Ridge (site 7), formed in coastal environments just before inundation of the land by the Western Interior Seaway in Cretaceous time. The striking high hogback just north of the Garden of the Gods park consists of Dakota Group sandstones.

Thanks to Christine Siddoway of Colorado College for her helpful comments.

The Lyons Formation thrusted westward over Fountain Formation at the north end of North Gateway Rock. The fault trace angles diagonally up toward the right from the arrow.
—Marli B. Miller photo

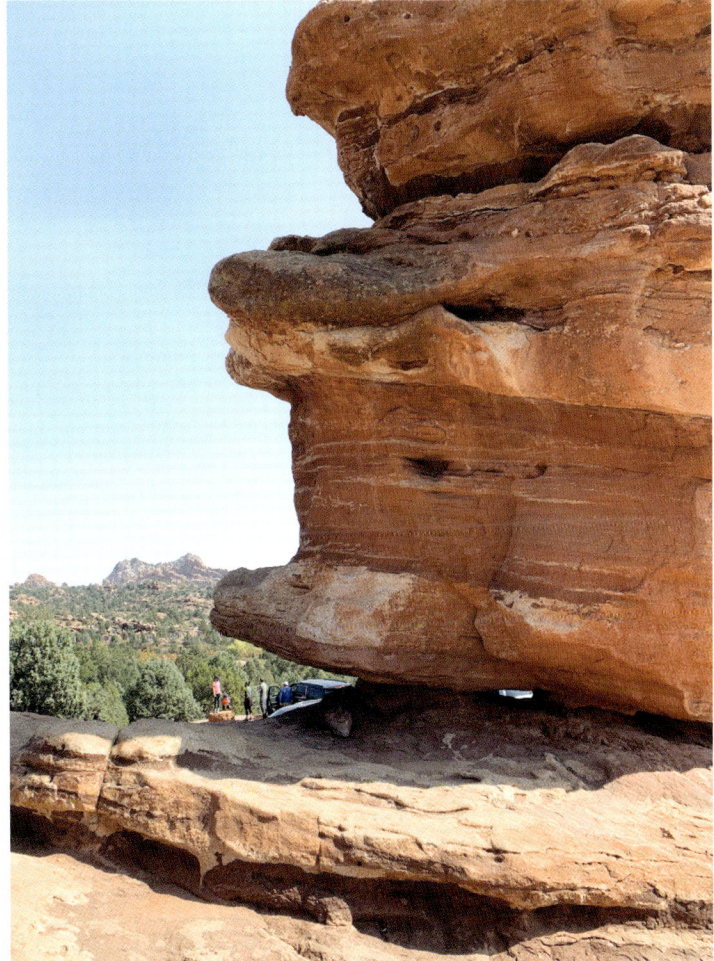

Balanced Rock, near the southwest corner of the park, is composed of the Fountain Formation. —Marli B. Miller photo

11 Paint Mines Interpretive Park
Badlands in Colorful Clays

Paint Mines Interpretive Park, 35 miles east of downtown Colorado Springs off US 24, is a hidden gem. The rolling topography at the parking lot is deceptively bland, but a short walk along the trail will bring you to a slope from where you can see a maze of towers, hoodoos, fins, and spires. This badlands landscape is constantly undergoing change because every rain or windstorm further erodes and shapes the bare sediment. This badland-style erosion reveals brilliantly colored sandstones and clays of the Pierre Shale, Fox Hills Sandstone, Laramie Formation, and Dawson Arkose along with a few instances of the Castle Rock Conglomerate.

The Paint Mines are located in the hummocky topography of the eastern edge of the Denver Basin, a downwarp at the edge of the Front Range. The sediments that fill the Denver Basin were eroded from the rising Rocky Mountains to the west and carried east by rivers in Late Cretaceous to early Cenozoic time in a much warmer and wetter climate than present. Sandstones were deposited along the shore of the receding Western Interior Seaway, while the clay of the Pierre Shale was deposited at the bottom of the ocean. Dark-gray clay layers of the Laramie Formation may preserve leaf fossils or imprints, indicating a swamp-like environment at the time of deposition. The easily eroded clays and soft sandstones form the striking towers, capped with more resistant Dawson Arkose or Castle Rock Conglomerate that protects the underlying layers from erosion. The Dawson Arkose contains clasts eroded from the Pikes Peak batholith, visible in the mountains on the western horizon. The red and purple coloring comes from the oxidation of iron minerals within the units; the white layers are largely quartz sandstones. Native Americans used the colorful clay for paint and pottery.

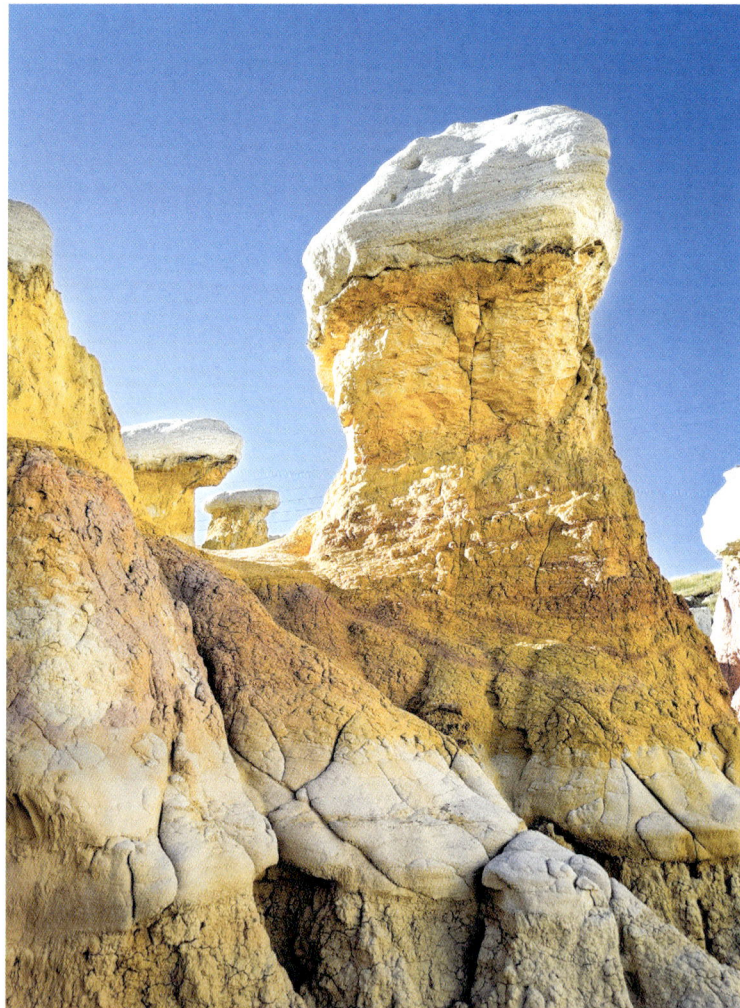

Hoodoos stand brightly against the sky in Paint Mines Interpretive Park. Cross-bedded white sandstone protects the towers of brightly hued mudstone beneath it. Eventually, erosion will narrow and undercut the caprock to the point that it topples, leaving the spires unprotected and eroding. —Magdalena S. Donahue photo

To reach the site from Calhan on US 24, turn south on Yoder Street (North Calhan Highway) on the east side of town. Drive 0.7 mile south and turn left (east) on Paint Mine Road and continue another 1.4 miles, following signage.

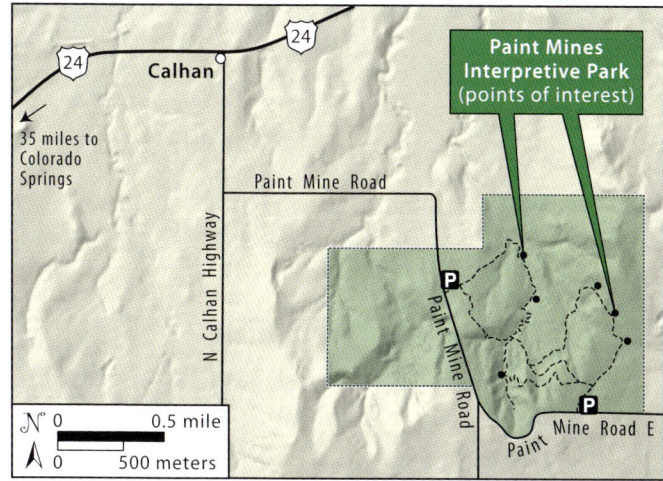

Paint Mines Interpretive Park
(points of interest)

24
Calhan
24

35 miles to Colorado Springs

Paint Mine Road

N Calhan Highway

Paint Mine Road

P

P

Paint Mine Road E

Paint

N
0   0.5 mile
0   500 meters

White rocks erode out of the surrounding hillslopes, showcasing spectacular shapes. —Magdalena S. Donahue photo

12 Cave of the Winds
Remaking of the Manitou Limestone

Cave of the Winds, west of Colorado Springs on US 24, offers a theme-park atmosphere on the outside, but numerous caves within the mountain amuse at an entirely different level. The caves are in the Manitou Limestone, a carbonate rock deposited in a shallow sea about 500 to 450 million years ago in later Cambrian to Ordovician time. The limestone contains trilobite and brachiopod fossils. For several hundred million years following deposition, the Manitou was gradually buried under more and more sediment, hidden from sight and from the erosive action of water.

Water first percolated through the rock formation beginning in the Laramide orogeny, the period of intense folding, uplift, and mountain building 75 to 40 million years ago. As major faults formed during the uplift, groundwater reached the Manitou Limestone along fractures. Cave formation truly began when the groundwater level dropped below the level of the Manitou Limestone approximately 7 to 4 million years ago. This enabled rainwater mixed with carbon dioxide to form a weak carbonic acid, to infiltrate and flow through the fractured rock. This acidic water rapidly accelerated the erosion of the limestone, expanding the cracks into pockets, chambers, and giant caves.

While much of the initial cave formation involved dissolving and removing material to form voids in the rock, the delightful cave formations, called speleothems, are new material that was deposited after the caves formed. As water percolates through overlying rock, it drips from low points in the cave roof, depositing microscopic amounts of calcium on cave ceilings and floors. Stalagmites grow up from the floor and stalactites grow down from the ceiling. Long, spectacular draping formations of travertine, called flowstone, formed as calcium carbonate was incrementally deposited from thin films or sheets of water flowing gently over the cave walls for millions of years. (See map with site 13 for directions to the cave.)

Flowstone in Cave of the Winds. —James St. John photo

Great unconformity (arrow), where the Cambrian-age Sawatch Sandstone (whitish band and reddish rock above band) overlies 1.04-billion-year-old Pikes Peak Granite, as seen looking northwest from the edge of the Cave of the Winds parking lot. —Marli B. Miller photo

View up Williams Canyon toward the main entrance to Cave of the Winds. The Manitou Limestone forms the tan-colored cliffs. —Marli B. Miller photo

37

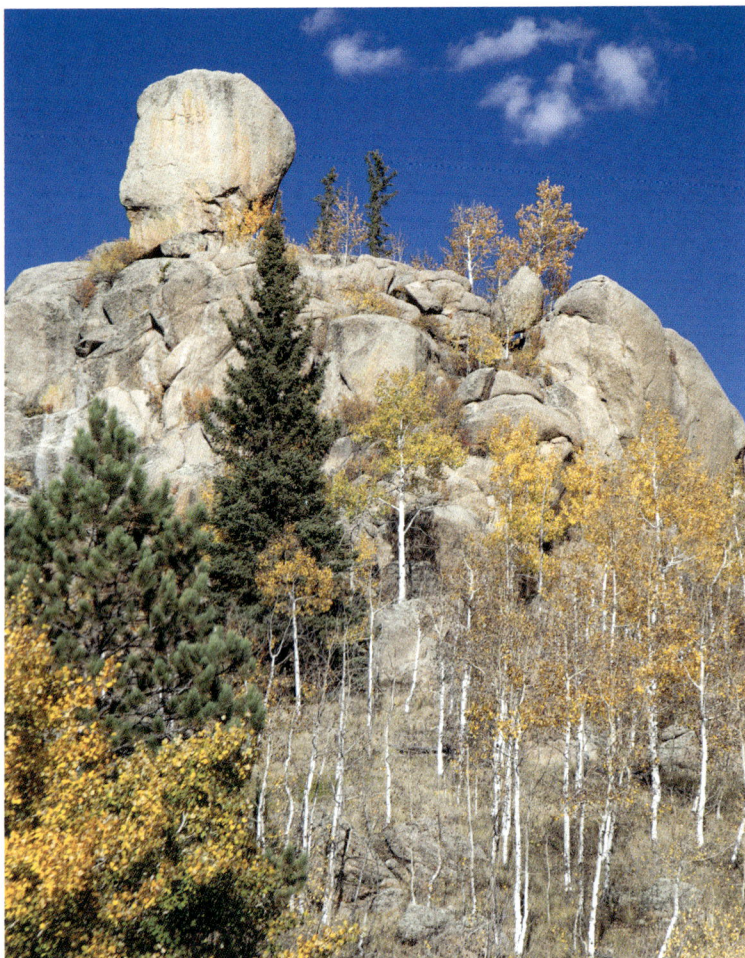

Pikes Peak Granite is an extremely uniform rock that weathers into rounded boulders. —Marli B. Miller photo

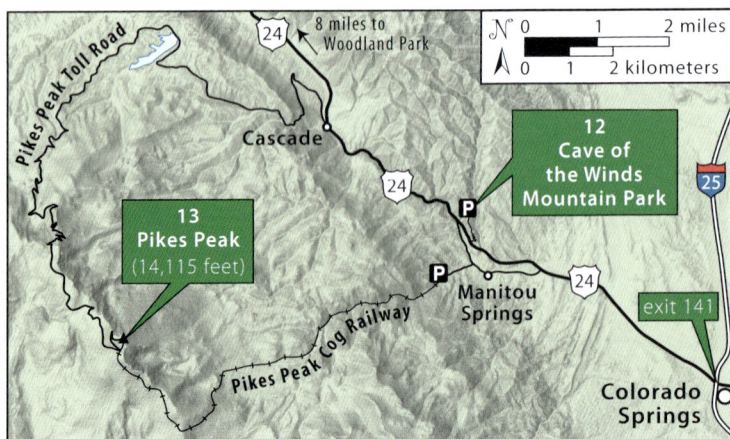

Pikes Peak
Pink Granite of a Fourteener

Pikes Peak, the crown jewel of the Front Range skyline near Colorado Springs, stands at 14,115 feet elevation, more than 8,000 feet higher than the city. The mountain is named for early explorer Zebulon Pike, who led an exploration expedition in 1806–1807 to investigate the headwaters of the Arkansas River in the Rocky Mountains. Pikes Peak is one of only two Fourteeners in Colorado that can be reached by motorized transport (Mt. Evans is the other one).

The Pikes Peak batholith was emplaced approximately 3 to 5 miles below the surface about 1.04 billion years ago and has since been lifted to spectacular heights. The batholith covers a large area measuring 80 miles long and 25 miles wide. Outcrops of distinctive rounded boulders show up westward along US 24 to well past Divide (where the photo at the left was taken) and surround the Eocene lakebed at Florissant Fossil Beds (site 14). The batholith is composed of two major plutons, the more potassium-rich Pikes Peak Granite and a slightly younger more sodium-rich series of syenites and granites. Syenite is similar to granite but lacks the abundant quartz seen in granites. Pikes Peak is largely composed of the Pikes Peak Granite, a variably fine to coarsely crystalline, pinkish rock dominated by potassium-feldspar, quartz, and biotite mica. Its pink to red color comes from the large amount of potassium feldspar, often in the form of large, 1-inch-long rectangular crystals.

Pikes Peak looms over Colorado Springs. While you can summit the peak on foot on a trail or in your own car on the toll road, it is easier to ride the Pikes Peak Cog Railway from Manitou Springs or take a bus tour to the top. To get to Cave of the Winds Mountain Park where you can take a cave tour, turn north from US 24 on Cave of the Winds Road on the east side of Manitou Springs.

The typical range of color in Pikes Peak Granite. Each photo is about 6 inches across. —Marli B. Miller photo

Pikes Peak as seen from the north near Woodland Park. The high, barren peak (at left) and ridgeline (on right) all consist of Pikes Peak Granite.
—Marli B. Miller photo

14 Florissant Fossil Beds National Monument
Volcanic Destruction Preserves Evidence of Life

Approximately 34 million years ago, at the end of Eocene time, the future site of Florissant Fossil Beds National Monument witnessed great destruction but also great preservation. Fast-moving lahars—volcanic mudflows—from the stratovolcanoes of the Guffey volcanic complex in the nearby Thirtynine Mile volcanic field dammed an ancient stream to form a lake. Volcanic ash continued to fall on the new lake, and fine-grained muddy sediments settled to its bottom to become the lowest shales of the Florissant Formation. This many-layered unit, a series of shales and mudstones with interspersed volcanic flows and ashes, records the history of the lake and what lived (and died) in it.

Giant redwoods that grew along the lakeshore were buried by a lahar and are now preserved and visible within the monument. These ancient redwoods were similar to modern sequoias and are estimated to have been between 500 and 700 years old at the time of their burial, according

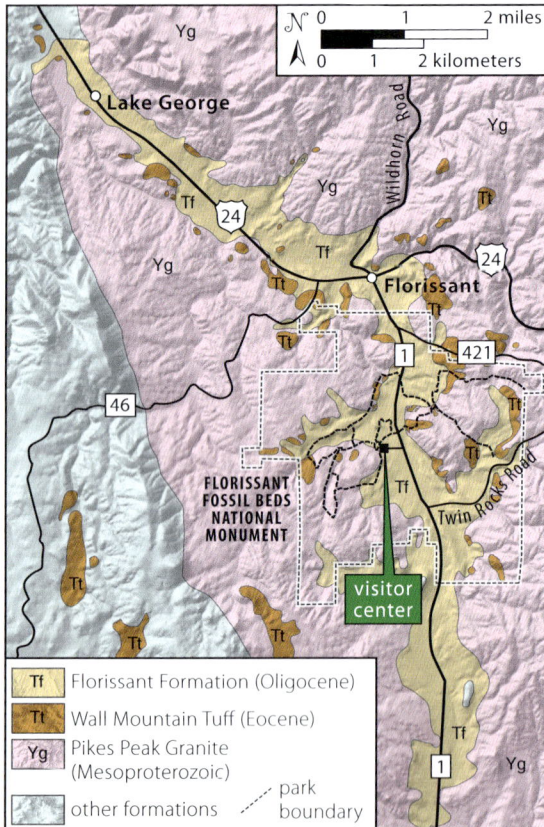

Tf	Florissant Formation (Oligocene)
Tt	Wall Mountain Tuff (Eocene)
Yg	Pikes Peak Granite (Mesoproterozoic)
	other formations

park boundary

Named a national monument in 1969, Florissant Fossil Beds is home to ongoing paleontological research. Visitors can explore the local geologic history along the monument's abundant hiking trails. The Florissant Formation shows the extent of the Florissant Lake that flooded the valley. Wall Mountain Tuff, a 36.7-million-year-old unit, forms big outcrops at the park.

A redwood branch was compressed and preserved between layers of sediment. —National Park Service photo

to tree ring analyses. These stumps were preserved by the process of permineralization, where organic material from the tree is replaced by silica and siliceous minerals as the tree is submerged in lake water, volcanic ash, and clay and mud layers. The texture of the woody stems is beautifully preserved in this process.

Other plant fossils include microscopic pollens, leaves, seed cones, and flowers that record a temperate and lush environment. Benthic diatoms are notable. Easy to preserve due to their silicic shell, this type of algae underwent enormous blooms during times of volcanism, when abundant silica ash was present in the lake waters. Following these blooms, the algal carpets died, sinking to the lake floor, where they blanketed the remains of other plants and animals, further helping to preserve them. Invertebrate fossils include arthropods, millipedes, insects, mollusks, and many arachnids. Vertebrate fossils are rarer, but include bottom-feeding fish, cuckoos, and shorebirds. Mammal, reptile, and amphibian fossils are nearly nonexistent in this location.

Tree rings and internal tree structures are beautifully preserved in this petrified tree stump. The original woody material was slowly replaced by silica and calcite. —Magdalena S. Donahue photo

Petrified stumps of redwoods of the warm, subtropical forest that once stood in this valley. —Magdalena S. Donahue photo

41

15 Royal Gorge
Grand Canyon of the Arkansas River

The Arkansas River exits the Front Range through the Royal Gorge, also known as the Grand Canyon of the Arkansas, but you wouldn't suspect the presence of this massive chasm from anywhere along US 50, the road that follows the Arkansas River between Pueblo and Salida. Hidden from sight from roads and towns, the canyon reaches 1,050 feet deep and is 1,250 feet across at its narrowest point upstream from Cañon City. You can view the deepest, narrowest section of the canyon from a suspension bridge nearly 1,000 feet above the river or from a railroad at the bottom of the canyon.

The canyon runs for nearly 6 miles, cutting through granite and gneiss bedrock of the Idaho Springs Formation, a complex of metamorphic rocks. Originally sedimentary and volcanic deposits, they were buried and metamorphosed 1.7 billion years ago. The Arkansas River began establishing its course through sedimentary rock as it drained the east side of the rising Rocky Mountains as they were uplifted during the Laramide orogeny. By the time the river reached the hard, crystalline bedrock, it had become confined to its course. As the landscape continued to move upward, the river cut down through the tough rock. The steep, rugged canyon walls exist in this location due to the relatively rapid incision of the river over the past 3 to 5 million years. Increased water flow during the Pleistocene ice ages also contributed to the carving of the canyon. Even today, the Arkansas River has extreme cutting power because the abundant waters of melting snow, rain, and glaciers combine with a large change in elevation. The landscape within its drainage basin plunges nearly 10,000 feet from snowy mountain peaks of the Mosquito and Sawatch Ranges to the plains to the east. Also contributing to the steepness of the canyon is the hard rock that maintains near vertical walls.

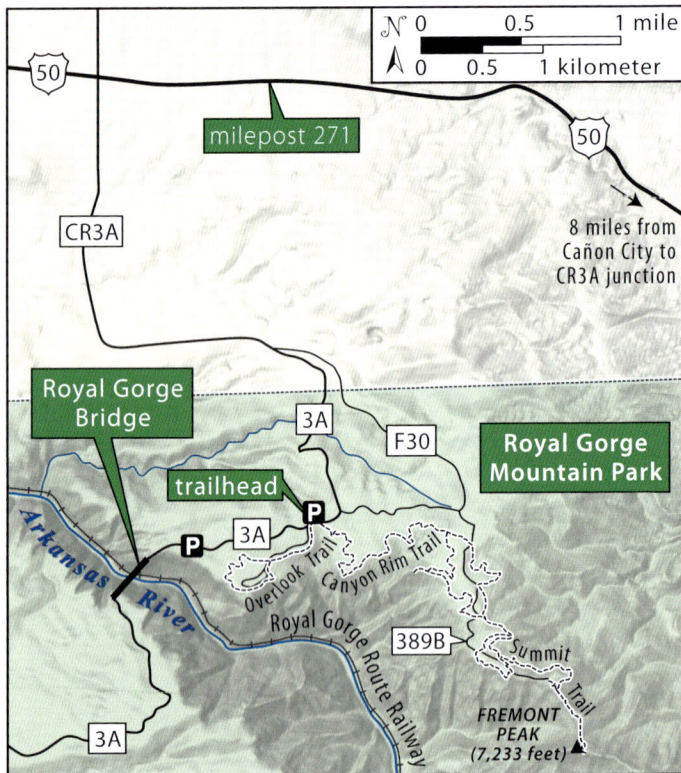

To view the canyon from the suspension bridge or railroad, you must first pay to enter Royal Gorge Bridge and Park, an amusement park. Other canyon views can be obtained, free of charge, by hiking the Overlook Trail or Canyon Rim Trail, just south of the Royal Gorge Bridge in the Royal Gorge Park Trail System.

View looking west at Royal Gorge from the western loop of the Overlook Trail Loop. Bands of lighter, more quartz-rich rock twist within the darker metamorphic rock, showing a long history of deformation and uplift. —Magdalena S. Donahue photo

16 Picketwire Canyonlands in Comanche National Grassland
Dinosaur Tracks along the Purgatoire River

The Purgatoire River track site, one of the largest dinosaur track locations in western North America, is both world famous and quite remote. After driving across the plains of southeast Colorado to the trailhead (two hours from Pueblo), you must then hike or mountain bike a 5.6-mile-long trail (11.3 miles round trip; bring lots of water on hot days) within the Comanche National Grassland. The site is also known as the Picketwire Canyonlands because Picketwire is a slang for Purgatoire, the river along which the tracks are found. You must cross the river to see the tracks.

More than 100 sets of track paths are preserved, including over 1,500 individual footprints. The dinosaurs walked through a wet, limy lakeshore with abundant vegetation, leaving the tracks in what is now limestone of the Morrison Formation of Late Jurassic age. The tracks include those from carnivorous biped theropods (allosaurs), sauropods (brontosaurs), and duck-billed ornithopods. The geometry of the tracks indicates that the dinosaurs traveled in groups of varying ages, demonstrating that these dinosaurs lived in social groups. In particular, the *Allosaurus* tracks were made by large and small animals, supporting the theory these dinosaurs hunted and scavenged in packs.

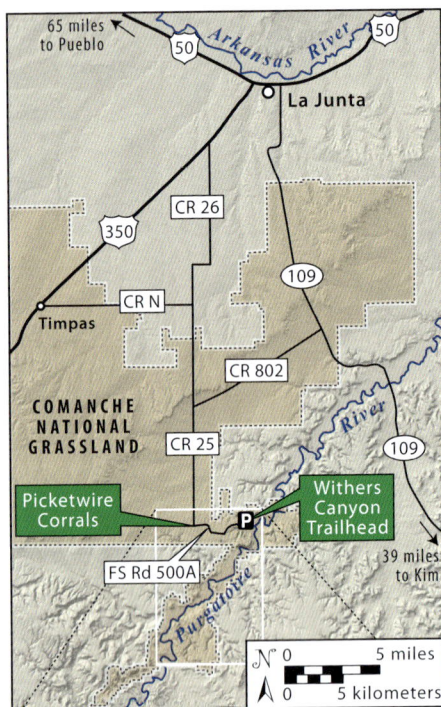

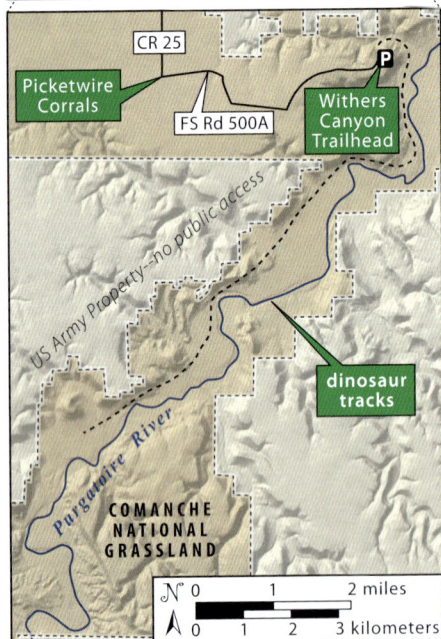

Water-filled tracks at Picketwire Canyon were left by a Brontosaurus as it trekked along an ancient lakeshore. The limy mud it sank into became a limestone bed in the Morrison Formation. —CM195902 photo, Creative Commons license 2.0

To reach Picketwire Canyon from La Junta, head 14 miles south on CO 109, then turn west (right) on a gravel road (David Canyon Road, County Road 802) and continue for about 17 miles to the Withers Canyon Trailhead.

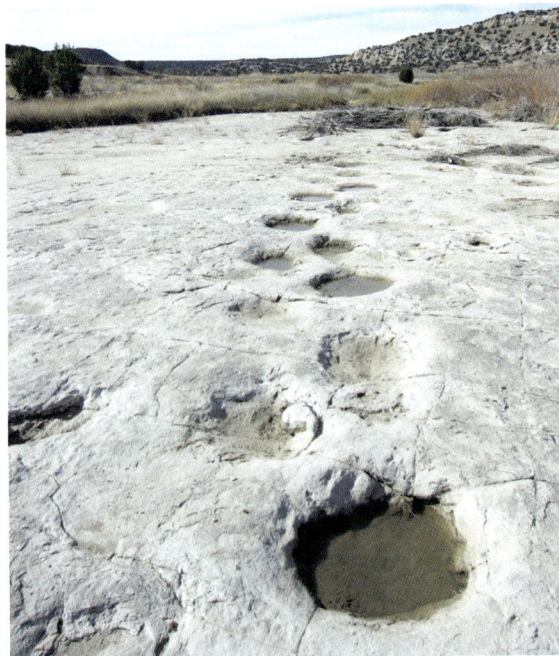

17 Trinidad Lake State Park
K-Pg Boundary Exposed

A thin rock layer in Trinidad Lake State Park records evidence of the catastrophic asteroid impact that resulted in the massive extinction of most non-avian dinosaurs 66 million years ago. This rock layer is at the Cretaceous-Paleogene (K-Pg) boundary, formerly known as the Cretaceous-Tertiary (K-T) boundary. (In a nomenclature move nearly as controversial as the boundary itself, the Tertiary Period was replaced in 2013 with the Paleogene and Neogene Periods.) The geologic time scale is based in part on changes in life, so major boundaries are placed at major extinction events. The K-Pg boundary marks the end of the Mesozoic Era (Age of Dinosaurs) and the beginning of the Cenozoic Era (Age of Mammals).

After the asteroid struck the Yucatán Peninsula of Mexico, nearly 75 percent of animal species on Earth went extinct, but they were not all killed instantly. Dust from the impact along with ash from wildfires reduced plant life enough that animals likely starved in the wake of the impact. The 1-centimeter-thick K-Pg boundary layer at Trinidad is full of unique materials, including high levels of iridium, shocked quartz, and microtektites. Iridium, the second-densest known metal, is found in much greater quantities in meteorites than on Earth. Natural shocked quartz can form only in response to an impact, and microtektites are round, pebble-like glass rocks that form during meteor impact. Scientists also observed a massive reduction of pollen and a surge in fern species after the time of impact. All this information is preserved in only a few inches of the rock record!

The rocks at Trinidad Lake State Park are part of the Raton Formation. Below the white boundary clay layer is a black, flaky coal layer formed in a swamp. Above the white boundary layer is another coal layer that some scientists believe formed in response to the global-scale wildfires and erosion that occurred post-impact. Above that are flaky brown mudstones and massive overlying sandstones of early Cenozoic time.

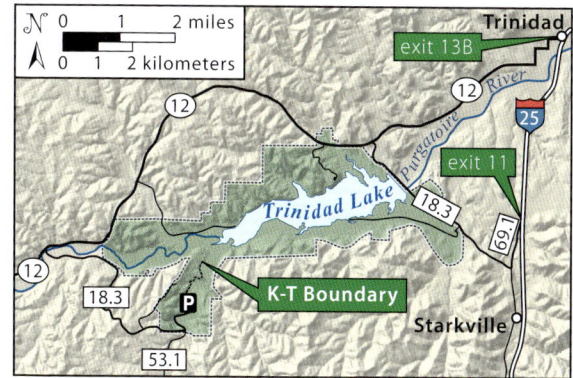

Look for the K-Pg layer along Longs Canyon Trail in the southwestern part of Trinidad Lake State Park.

The lower, thinner bright white layer in the Raton Formation contains evidence of the K-Pg impact. —Magdalena S. Donahue photo

18 Radial Dikes of the Spanish Peaks
Magma-Filled Fractures Form Vertical Walls

Hundreds of dikes extend outward from the Spanish Peaks like spokes on a wheel. It is not unusual for dikes to emanate from a central magmatic intrusion such as the Spanish Peaks, but it is unusual for them to be so prominent, standing as vertical walls that rise above the countryside. Except where baked hard by the magma, the less-resistant sedimentary rocks that once surrounded the magmatic intrusions have eroded away. The dikes stand nearly 100 feet high, and the Spanish Peaks tower more than 7,500 feet above the Great Plains, the western edge of which is only about 15 miles to the east.

The magmas that formed the Spanish Peaks, being hot and buoyant, intruded into the crust and cooled without much surface eruption. During these intrusions from about 24 million years ago, the overlying bedrock was warped upward, draping itself around the molten rock and cracking and fracturing in all directions. The magma took advantage of these cracks, squishing outward in a radial pattern from the Spanish Peaks center. These dikes range in size from 1 foot to more than 100 feet wide and run up to 14 miles long. The majority of the dikes tend to be straight and do not branch or intersect.

East and West Spanish Peaks consist of related but slightly different compositions. West Spanish Peak is slightly older quartz syenite, a plutonic rock rich in alkali and plagioclase feldspar minerals. East Spanish Peak is a slightly younger granodiorite porphyry, a similar type of plutonic rock but having more quartz and plagioclase feldspar. The magma chemistry changed as eruptions occurred and new magmas were generated. The magma was produced at the same time the Rio Grande rift was opening to the west.

A vertical dike stands above the landscape in the foreground with West Spanish Peak in distance. —Marli B. Miller photo

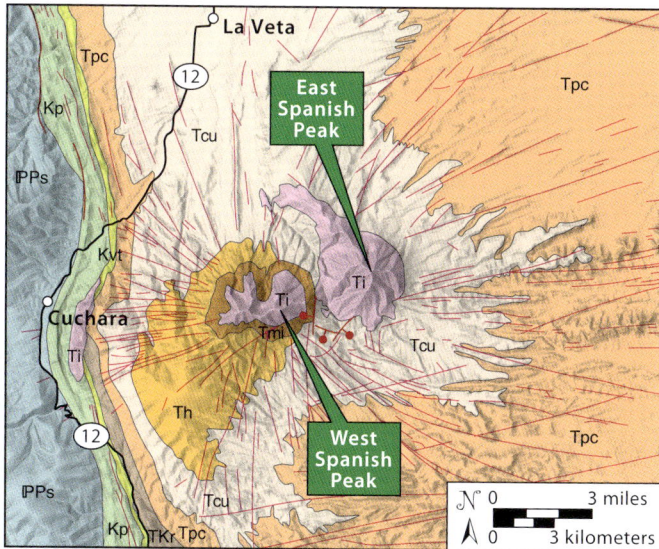

To see a dike up close, head south on CO 12 from La Veta.

CENOZOIC

- **Ti** intrusive rocks
- **Tmi** metamorphic rocks
- **Th** Huerfano Formation (Eocene)
- **Tcu** Cuchara Formation (Eocene)
- **Tpc** Poison Canyon Formation (Paleocene)
- **TKr** Raton Formation (Cretaceous-Paleocene)

dikes

normal fault

MESOZOIC

- **Kvt** Vermijo Formation (Cretaceous)
- **Kp** Pierre Shale; includes Dakota Group and Purgatoire Formation (Cretaceous)

PALEOZOIC

- **ℙPs** Sangre de Cristo Formation (Pennsylvanian-Permian)

Roadside exposure of the dike (light rock at right) intruding the Cuchara Formation (dark rock at left), Location is a few miles south of La Veta on CO 12. In the close-up, note the large feldspar crystals in the dike rock. —Marli B. Miller photo

47

CENTRAL ROCKIES

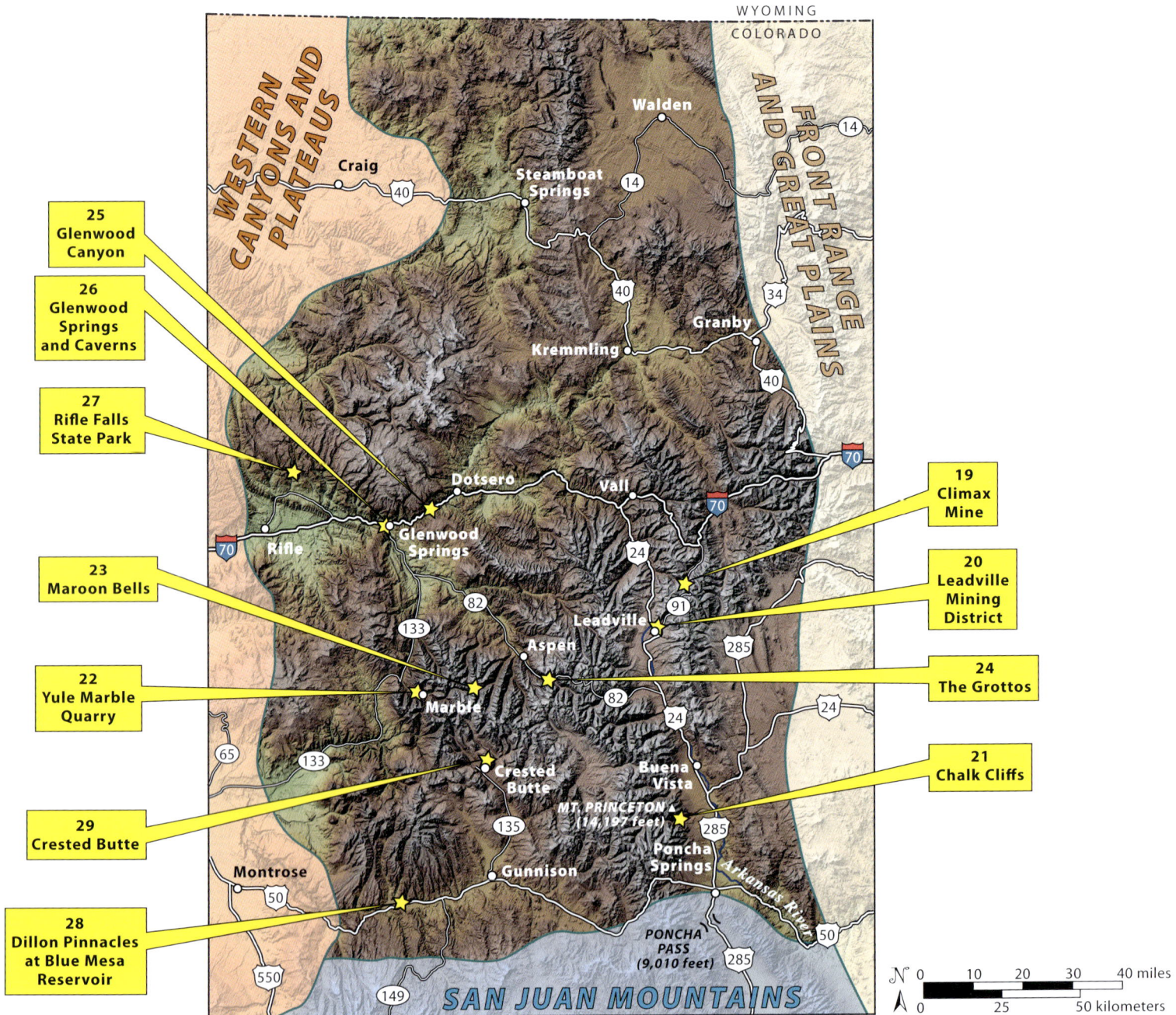

WYOMING
COLORADO

WESTERN CANYONS AND PLATEAUS

FRONT RANGE AND GREAT PLAINS

SAN JUAN MOUNTAINS

25 Glenwood Canyon

26 Glenwood Springs and Caverns

27 Rifle Falls State Park

23 Maroon Bells

22 Yule Marble Quarry

29 Crested Butte

28 Dillon Pinnacles at Blue Mesa Reservoir

19 Climax Mine

20 Leadville Mining District

24 The Grottos

21 Chalk Cliffs

Walden

Steamboat Springs

Craig

Granby

Kremmling

Vail

Dotsero

Glenwood Springs

Rifle

Leadville

Aspen

Marble

Crested Butte

Buena Vista

MT. PRINCETON ▲ (14,197 feet)

Poncha Springs

Gunnison

Montrose

PONCHA PASS (9,010 feet)

Arkansas River

14

40

14

40

34

40

40

70

70

24

82

133

91

285

82

24

24

135

285

285

65

133

50

550

149

50

N
0   10   20   30   40 miles
0   25   50 kilometers

West Elk Mountains looking north from Crested Butte. Note the jagged peaks and U-shaped valleys, the erosional hallmarks of alpine glaciers. —Magdalena S. Donahue photo

Long before peak baggers, hikers, mountain bikers, and river rafters began exploring the central Rocky Mountains, prospectors scoured the rugged landscape in search of gold and silver. They found rich mineral and ore deposits in the northeast-trending Colorado Mineral Belt, a 10- to 35-mile-wide feature extending 250 miles from near Durango to Boulder. The belt lies along a northeast-trending structure that formed 1.7 billion years ago in Proterozoic time as continents coalesced and sheared. Folding and metamorphism occurred in the zone during the Laramide orogeny, followed by magmatic intrusion and volcanism during the Oligocene Epoch. Laramide intrusions, Oligocene volcanism, and Rio Grande rifting were responsible for most of the valuable mineral deposits in the Colorado Mineral Belt. When hot magma intruded the crust, it brought with it superheated, mineral-saturated waters that moved through the already fractured and faulted bedrock. The area still has high heat flow, which we can see every day in the number of hot springs around the central Rocky Mountains.

19 Climax Mine
Molybdenum Mine on Continental Divide

The Climax Mine, at 11,360 feet on the Continental Divide near Leadville, is the world's largest open-pit molybdenum mine. It is likely also the world's highest and coldest, a difficult location to live and work, even with modern conveniences. Early mines in the region targeted gold and silver, which occurred with lead, zinc, and copper. Prospectors thought that the network of tiny dark veins found along fractures in a light-gray rock were lead, but the veins turned out to contain molybdenum, or "moly" in mining vernacular. It is present at the Climax locality as molybdenite, a low-grade sulfide mineral. While molybdenum has many uses today, it is best known for its ability to make hard, durable steel alloys, critical to weapons production. The Climax

Mine opened in 1915 as war raged in Europe, and later it supplied all the molybdenum for the United States and its allies in World War II. The mine has opened and shut multiple times as the global price of moly fluctuated. The Climax Mine also produced tin, tungsten, and pyrite.

Moly deposition at the Climax Mine occurred during the intrusion of seven igneous bodies during the Oligocene Epoch. These magmatic intrusions invaded Proterozoic schist and granite and cooled there. One of these intrusions, known as the Climax Stock, is composed largely of rhyolite and granite with a variety of crystal sizes, indicating multiple stages of cooling of the magma chamber. How quickly magmas cool affects the size of minerals we see: slow-cooling

Tiers of rock rise above mining facilities at the Climax Mine, which straddles the Continental Divide at Fremont Pass. —Magdalena S. Donahue photo

magma has time to form large crystals, while more rapid cooling results in tiny to microscopic crystals that form the matrix, or groundmass, of the rock.

Originally mined underground, the ore is mined today in a large open pit due to the low-grade nature of the ore deposit. More than 2,000 pounds of material must be mined, milled, and processed to produce 6 pounds of molybdenum! Large molybdenum deposits have also been mined near the towns of Empire, Colorado, and Questa, New Mexico. You can easily see the mine and its tailings piles about a dozen miles northeast of Leadville along CO 91 at the top of Fremont Pass.

A cross-section of molybdenite-quartz veins in granite. Dark silvery gray molybdenite stands out against the medium-gray quartz and light-colored granite. —James St. John photo

After molybdenum is chemically removed from the crushed rock, the water slurry is left in tailings ponds. —Marli B. Miller photo

Abandoned housing near in the Gilman-Leadville mining area shows the rugged topography in which miners lived and worked.
—Magdalena S. Donahue photo

20 Leadville Mining District
Silver Ore in Dolomite

The Leadville mining district, one of the richest mining districts in the world, lies in the center of the Colorado Mineral Belt. Most of the sulfide mineral ore, deposited after the end of the Laramide orogeny, is in tiny microveins in the highly fractured rock. The ore minerals include pyrite, galena, sphalerite, chalcopyrite, gold, silver, and copper. The silver that the mining district is known for is found in veins associated with manganese and lead within three dolomitic geologic formations: the Leadville Limestone, Dyer Dolomite, and Manitou Limestone. Though not intrinsically valuable, pyrite is important for miners because it can indicate that gold is nearby. Pyrite forms under the same conditions and in the same rock types as gold does.

Dolomites readily host ore deposits. Dolomite forms where calcium carbonate of a limestone deposit is replaced with magnesium carbonate. Because magnesium has a

Pyrite from the Gilman mining area. —James St. John photo

smaller atomic size than calcium, minute pockets of space are present after limestone is converted to dolomite. Mineral- and metal-saturated waters circulating through the pores deposit metals within these small, interconnected cavities as the water cools.

The Leadville mining district was first developed following the discovery of gold in stream deposits in nearby Oro City and California Gulch in 1860. Miners discovered the mineral cerussite, a lead mineral high in silver, in the placer sands in Oro City in 1876. Miners searched upstream, looking for and finding the ore deposits that the heavy sands eroded from. The city of Leadville was established in 1877, leading the Colorado silver boom. Among notable mines in the district is Sherman Mine, the most productive, and Black Cloud Mine, the longest lived, closing in 1999. The environmental impacts of such a long history of mining are not inconsequential. Dissolving, or leaching, thousands of tons of rock to extract the desired particles of ore contaminated surface and groundwater with arsenic and lead. Following years of work, the Environmental Protection Agency's mitigation effort has largely been successful.

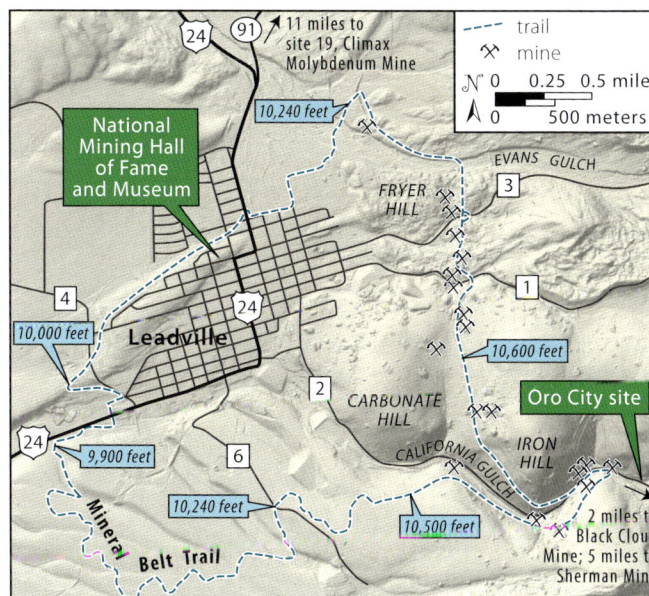

You can see much of the historic mining area along the 11.6-mile paved Mineral Belt Trail that circles through the Leadville area. Mineral samples and additional information about the people who built their fortunes on the region's mining can be seen at the National Mining Hall of Fame and Museum in Leadville.

53

21 Chalk Cliffs of Mt. Princeton
Rock Altered by Hot Springs along Rift Fault

Mt. Princeton, one of the most dramatic peaks in the Sawatch Range at 14,196 feet, rises nearly 7,000 feet above the Arkansas River valley. At the base of the mountain, near Mt. Princeton Hot Springs, are white chalky cliffs, the formation of which are intertwined with the formation of the valley. The Arkansas River valley is near the northern end of the Rio Grande rift in which the Earth's crust from Colorado south through New Mexico was split, or pulled apart. To accommodate the stretching, blocks of the crust drop down along faults. Rifting was mainly active in this region from approximately 28 to 12 million years ago, and the entire length of the Rio Grande rift was pulled apart nearly at the same time. According to many researchers, the central Rio Grande rift is still deforming, although likely at a slower rate than earlier.

The Arkansas Valley is an asymmetrical basin with the western side dropped down along the Sawatch fault. Mt. Princeton has been lifted up along the west side of the fault. The mountain is composed of the Mt. Princeton batholith, a large quartz monzonite body that cooled at relatively shallow crustal depth between approximately 35 and 30 million years ago. The Sawatch fault exposes a huge section of the batholith.

Aerial view of Chalk Cliffs and fault-bounded Collegiate Peaks of the Sawatch Range, looking north. The rift valley extends into the distance along the right side of the photo.
—Marli B. Miller photo

The chalky cliffs of Mt. Princeton are visible in the flank of the mountain, behind irrigated agricultural fields.
—Magdalena S. Donahue photo

Rocks below the surface here are hot because of the thinning of the crust and rising magma associated with the rift. Heated groundwater rises along faults, so several hot springs exist along the fault at the base of Mt. Princeton, the most prominent of which are the Mt. Princeton and Hortense Hot Springs. The Hortense Hot Spring has recorded temperatures of 183 degrees Fahrenheit, making it the hottest hot spring in Colorado! Hydrothermal activity associated with the hot springs altered parts of the batholith rock to laumontite, a white zeolite mineral. Rising nearly 1,600 feet above the valley floor, the white Chalk Cliffs get their coloring from the laumontite that fills the fractured rock, brilliantly outshining the gray host rock.

The astute reader will wonder why the Arkansas River doesn't flow into the Rio Grande if the valley here is part of the Rio Grande rift. Geologists think that it did once, but volcanism, uplift, and associated faulting near Poncha Pass forced the Arkansas River to turn east at Salida.

22 Yule Marble Quarry
Pearly White Building Stone

The uniform, pearly white stone of the Lincoln Memorial in Washington DC was quarried from a remote area in Colorado's Elk Mountains. Marble was first discovered in 1873 in the Yule Creek valley by geologist Sylvester Richardson. The Yule Marble is named after prospector George Yule, who rediscovered the marble in 1874 after the original findings failed to garner interest among prospectors more focused on gold, silver, and other precious metals.

The Yule Marble formed when hot magmas intruded into the Leadville Limestone and crystallized deep underground into granite. Contact metamorphism, the heating of the bedrock surrounding a hot magma body, rearranged and recrystallized the limestone into 99.5 percent pure calcium carbonate with minor tiny quartz inclusions and a beautiful grain structure. Contact metamorphism results in a very pure and consistent marble, unlike many other marbles that form during periods of regional metamorphism associated with the construction of huge mountain belts. The marble and granite were brought to the surface as the Rocky Mountains were uplifted, and the granite now forms nearby Treasure Mountain dome.

The Yule marble began to be quarried and exported from Colorado in 1884 by the Colorado Marble and Mining Company. The first quarry workers faced many of the same challenges that miners looking for ore and precious metals did: elevation between 9,000 feet and 10,000 feet, extremely steep and rugged topography, punishing winters, lack of mining and transportation infrastructure, and limited access to supplies. These difficulties resulted in several company turnovers before the Colorado-Yule Marble Company developed the Yule Marble Quarry, which is still the location of today's marble production near the town of Marble. In addition to the Lincoln Memorial, this marble has been used in the Tomb of the Unknown Soldier and George Washington Monument in DC and several prominent buildings in Colorado including the Colorado State Capitol Building, the Federal Reserve Bank, the Colorado State Museum, and the Denver Post Office.

Today's mining operations take place along a narrow band that follows the inclined edge of the Leadville Limestone, metamorphosed to marble by a nearby granite intrusion.
—Marli B. Miller photo

Close-up view of the unusually pure white marble. —Marli B. Miller photo

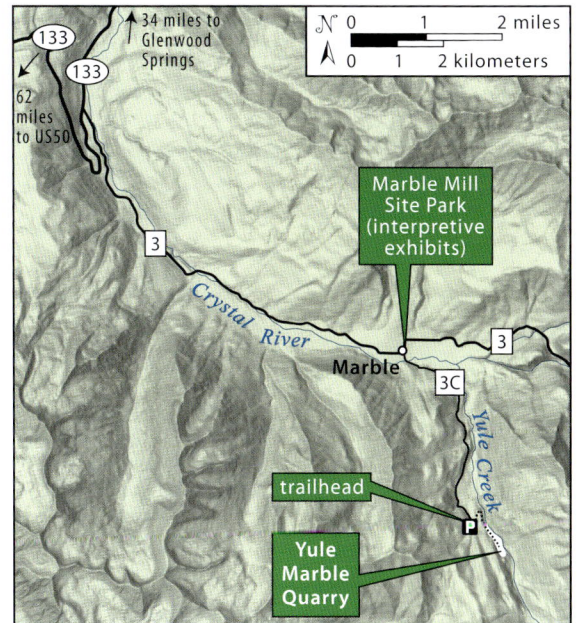

Workers with marble in 1918. Workers drilled holes to break the rock into blocks and used feather wedges to further loosen each stone block. —Library of Congress

Blocks of unused or low-quality marble sit near the quarry. —Marli B. Miller photo

You can reach the town of Marble by heading about 22 miles south of Carbondale on CO 133, then turning south onto County Road 3 and driving another 5.5 miles to the town of Marble. The Yule Marble Quarry presently offers no direct access or tours due to safety regulations. However, you can see some of the operations from a distance if you drive 3 miles up Forest Service Road 3C. Scrap stone can be seen and purchased when contacting the owning company, Colorado Stone Quarries, Inc.

23 Maroon Bells
Ringing High above Aspen

Two Colorado Fourteeners, Maroon Peak and North Maroon Peak, are collectively known as the Maroon Bells and rise high above the quaking aspen forests below. While many of the peaks in Colorado are made of granite and gneiss, the Maroon Bells in the Elk Mountains are composed of distinctively layered sedimentary rock—reddish feldspar-rich sandstone, conglomerate, mudstone, and minor limestone of the Maroon Formation. This unit is more than 10,000 feet thick. The sediments were originally eroded from the rising Ancestral Rockies and deposited in nearby basin floodplains and shorelines in Pennsylvanian to Permian time. Later, the sediments were cemented and hardened into rocks, then uplifted during the Laramide orogeny. Like the surrounding Rockies, the Maroon Bells were carved by glaciers, which also carved out the cirque in which nestles the much-photographed Maroon Lake. This iconic location, 11 miles southwest of Aspen, is extremely popular and open seasonally; visitation to Maroon Lake requires a reservation and fee for both personal car and shuttled trips.

Aspen, which is at the intersection of the Colorado Mineral Belt and the Sawatch uplift of Laramide age, began as a mining town in the 1870s, producing silver, lead, zinc, and minor gold. The Aspen mining district, also known as the Roaring Fork district, was one of the most productive silver districts in history, and its famous Smuggler Mine yielded the heaviest silver nugget ever found: 2,054 pounds with 93 percent purity. The Smuggler Mine produced ore from discovery in 1879 until 1917, when it closed for a while before reopening in 1970. The area was declared a Superfund Site in 1984, and though mining is ongoing, it is minor, and the mine is also used for tourism.

Early prospectors scoured the mountains around Aspen looking for the next big strike, but thankfully the Maroon Bells do not possess mineral ores. Although multiple intrusions cut through the rocks near the Maroon Bells and have even turned some of the rocks green through contact metamorphism, the magmatic fluids were mostly devoid of valuable metals. Many rocks contain the greenish, foliated, pearly-looking mineral chlorite, which forms during low-temperature metamorphism.

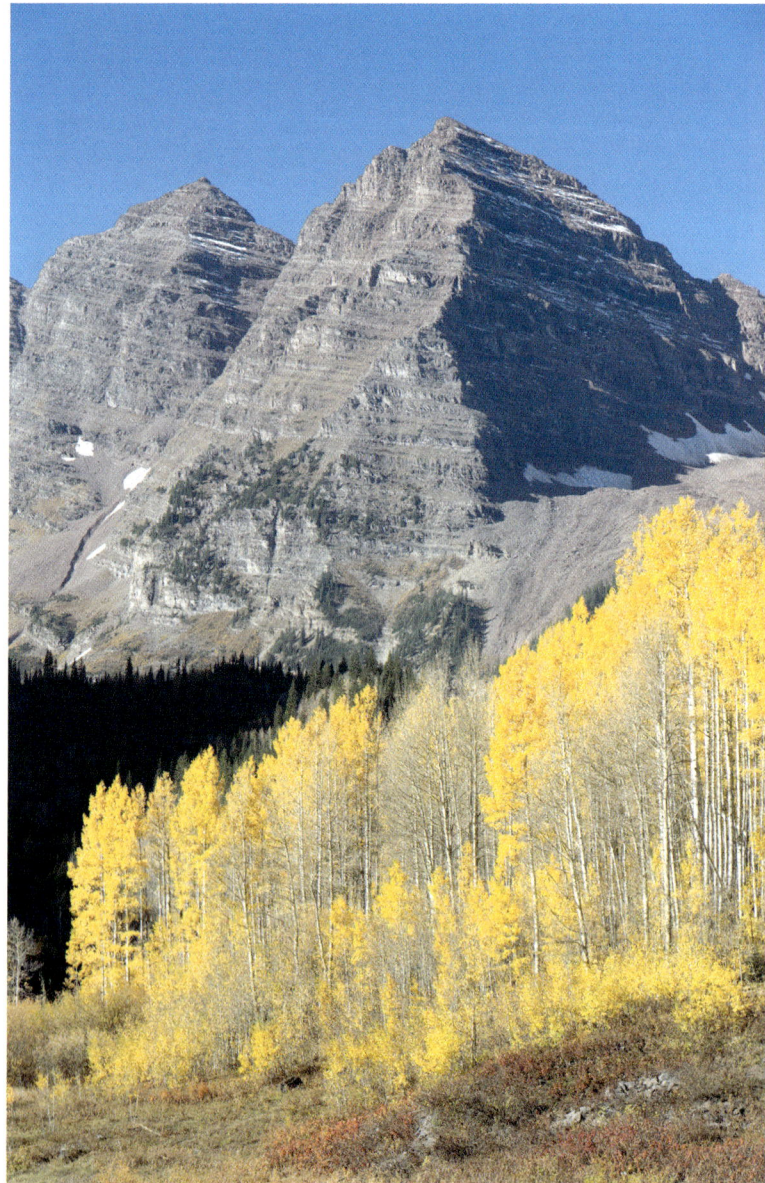

Dipping sedimentary beds of the Maroon Formation in the Maroon Bells, as seen from Maroon Lake. —Marli B. Miller photo

View from a few miles above Maroon Lake looking down the U-shaped glacial valley of West Maroon Creek toward Maroon Lake, which lies just past the bend in the valley. The Maroon Formation forms both the greenish and reddish cliffs on the left side of the valley. —Marli B. Miller photo

Three different rock types common to the Maroon Formation. Left: thinly bedded sandstone showing low-angle cross beds and erosional scours. Middle: Conglomerate and sandstone. Note how the conglomerate contains a variety of pebbles and cobbles, including quartzite, limestone, and Proterozoic basement. Right: green and tan-colored sandstone. The green color derives mostly from the mineral chlorite that formed during contact metamorphism. —Marli B. Miller photo

59

A glacial erratic sits on the glacially polished, undulating whaleback. —Marli B. Miller photo

24 The Grottos
Channels Eroded by a Subglacial Stream

Ever wonder how bedrock erodes beneath a glacier? It depends whether or not there's a stream flowing beneath the base of the glacier. In most places, rocks embedded in the base of the glacier scour and sometimes polish or groove the bedrock. In a few places, a stream flows between the bedrock and the ice. Subglacial streams can carve deep channels into the bedrock.

At the Grottos, along the Roaring Fork River south of Aspen, you see the effects of both these processes along a half-mile loop trail. Much of the trail follows an ice-scoured surface of the 1.4-billion-year-old granitic Grottos pluton.

Called a whaleback because of its smoothed, streamlined surface, you can see broad undulations abraded into the rock by other rocks embedded in the ice. You can also see glacial erratics, large boulders that were carried in the ice and then left behind when it melted. The Grottos, found near the northeast corner of the trail, consist of some labyrinthine passages carved by a subglacial river. Be especially careful about slick surfaces if you choose to explore them. To complete the loop, you can walk to the Cascades at the edge of the Roaring Fork River and see a smaller-scale version of the grottos being carved today.

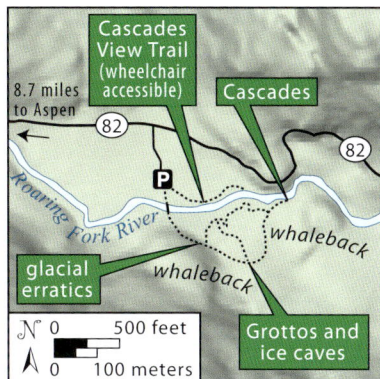

The Grottos are 9 miles southeast of Aspen along CO 82.

The Grottos were eroded into Proterozoic granite by a stream flowing beneath a glacier. —Marli B. Miller photo

The Roaring Fork River is eroding modern potholes at the Cascades. —Marli B. Miller photo

25 Glenwood Canyon
Engineering Feats near the Great Unconformity

With walls towering 1,300 feet above the Colorado River, Glenwood Canyon is the deepest canyon of the Colorado River in its Rocky Mountain headwaters. Draining much of western Colorado, the river originates north of Granby on the east side of Rocky Mountain National Park and flows 1,450 miles en route to the Gulf of California. At Glenwood Canyon, it is still relatively small but flows with considerable force.

Although Glenwood Canyon exposes Proterozoic basement rocks, the canyon itself is a relatively young feature. The Colorado River began carving it during the Pleistocene Epoch when a cooler climate brought increased precipitation, and runoff from the highlands was flush with glacial meltwater. The Leadville Limestone, a 360-million-year-old sedimentary rock, is the most visible unit as you follow the river through the canyon on I-70. The Leadville Limestone

The two Hanging Lake Tunnels of Interstate 70 emerge from the sheer walls of Glenwood Canyon. Each of these tunnels is 4,000 feet long, bored through the Leadville Limestone.
—Magdalena S. Donahue photo

The Great Unconformity, the contact between the layered Cambrian sedimentary rock and the massive, dike-riddled Proterozoic gneiss, is visible high above the canyon from near the Hanging Lake Rest Area. – Marli B. Miller photo

is bluish- to brownish-gray with sandy layers near the base. The upper part is mostly clastic limestone—carbonate sands cemented by calcite—with interbeds of reddish shale.

The Great Unconformity, where Paleozoic sedimentary rocks were deposited on an eroded surface of Proterozoic rock, makes an appearance in the lower canyon. Look for it from the Hanging Lake Rest Area, where the pinks of the Proterozoic basement rocks come against the orderly layers of the overlying, buff-colored Sawatch Sandstone of Cambrian age.

Glenwood Canyon is also the home to some of the greatest engineering in modern transportation. Interstate 70 runs west across the state from Denver, and the 12-mile section through the Glenwood Canyon was the last part built. Construction of this section spanned twelve years (1980–1992) and reached $490 million in cost. The engineering feats include bridges, viaducts, three tunnels, and fifteen miles of retaining walls. Significant portions of the highway are cantilevered, and in places the road lanes extend over the Colorado River. The canyon also hosts the Union Pacific Railroad, which is still in use.

The Hanging Lake Tunnels get their name from Hanging Lake, one of the most popular hiking destinations in Colorado. Travertine, a form of limestone, is continuously being deposited at the lake, creating a fragile and intricate lakeshore and walls that are slowly but constantly changing. Rainwater and groundwater percolating through faults and cracks in the Leadville Limestone dissolve calcium and carbonate minerals from the bedrock. Once the calcium carbonate reaches the surface via a spring, lower atmospheric pressure causes it to precipitate, or come out of solution, as tan, buff, or white deposits of travertine. The lake also hosts a unique hanging-garden plant ecosystem, including rare garden sullivantia, red-osier dogwood, river birch, bog orchid, and American speedwell. This protected National Natural Landmark receives heavy foot traffic and requires a permit and shuttle transport from Glenwood Springs. The trail was closed after fires raged through the canyon in 2020, so check for updated information.

At river level in Glenwood Canyon bedding layers of the Leadville Limestone are tilted. Water through Glenwood Canyon is highly regulated by upstream dams, and previous years' high-water marks are visible as the dark stripe above the current water level. Long black bacteria-tinted stains stripe the canyon's upper walls, where microorganisms thrive in water seeps. —Magdalena S. Donahue photo

26 Glenwood Springs and Caverns
A Network of Water and Rock

The town of Glenwood Springs, a popular recreational stop, sits at the confluence of the Roaring Fork and Colorado Rivers. The flowing water and rock walls are obvious at the surface, but it's what they do underground that makes it interesting. The Mississippian-age Leadville Limestone, which forms the steep, dipping walls on the north side of Glenwood Springs, is riddled with caves, including Glenwood Caverns and Fairy Cave. Slightly acidic water can rapidly erode limestone, particularly along cracks or fractures. As water circulation persists, the cracks widen into caves. Tectonic deformation associated with the raising of the Rocky Mountains has folded, faulted, and fractured the Leadville Limestone, so it was inevitable that water would form large cave networks. Cave tours are available as part of Glenwood Caverns Adventure Park.

As groundwater circulates upward through the fractured and faulted bedrock, it gathers heat and minerals. A collection of about ten to twelve hot springs forms the largest hot spring in Colorado, Glenwood Hot Springs. Their combined output flows at a rate of over 2,500 gallons per minute, and

temperatures reach 125 degrees Fahrenheit. The water contains abundant sulfur that gives the springs a distinctive rotten-egg odor. The spring water also contains a lot of salt because the groundwater dissolves halite (rock salt) and gypsum (calcium sulfate) from the Eagle Valley Evaporite. These springs contribute so much saline water to the Colorado River that water quality improvement projects include desalinization proposals!

Early indigenous people called the springs *Yampah*, meaning "Big Medicine," and these mineral-rich waters are still sought after for soaking and health benefits. The Glenwood Springs Resort, originally established in 1890, has an excellent selection of historical photos, including one near its large mother spring showing a historic advertisement for a hot spring water cocktail.

This historic recipe, posted at the Glenwood Springs Resort, touts the health benefits of drinking hot spring water. Please don't drink the water! —Magdalena S. Donahue photo

Glenwood Hot Springs Historic Drinking Spring

Here's to Your Health!

It's been decades since the mineral water was bottled, but many have continued the tradition of taking the original Glenwood Springs "cocktail."

Caution – Spring water is 122° and could cause burns.

GLENWOOD HOT SPRINGS RESORT

Swim for Health and Beauty

in the World's Largest

HOT MINERAL WATER SWIMMING POOL

Water Analysis

Azure-Yampah Spring, By Charles F. Chandler, Ph.D., New York

GRAINS IN ONE U. S. GALLON OF 231 INCHES OF WATER

Chloride of Sodium	1089.8307
Chloride of Magnesium	13.0994
Bromide of Sodium	0.5635
Iodide of Sodium	Trace
Fluoride of Calcium	Trace
Sulphate of Potassa	24.0434
Sulphate of Lime	82.3861
Bicarbonate of Lithia	0.2209
Bicarbonate of Magnesia	13.5522
Bicarbonate of Lime	24.5737
Bicarbonate of Iron	Trace
Phosphate of Soda	Trace
Biborate of Soda	Trace
Alumina	Trace
Silica	1.9712
Organic Matter	Trace
Total Solids	1250.0401

27 Rifle Falls State Park
Travertine Dam below the Leadville Limestone

Along the dry western edge of the Rockies north of Rifle is an unexpectedly lush area along East Rifle Creek. A three-pronged waterfall cascades nearly 70 feet over a cliff of travertine, a carbonate rock, at Rifle Falls State Park, located 14 miles north of the town of Rifle. Another body of travertine exists about 1 mile upstream (north) from the falls near the Rifle fault, which separates Pennsylvanian and younger bedrock on the south from older rock on the north. This older rock includes the Mississippian Leadville Limestone, which walls the canyon of East Rifle Creek and provides much of the calcium carbonate that is re-precipitated as travertine.

As the calcium-rich river and groundwater lose dissolved carbon dioxide to the atmosphere, particularly at the churning falls, it deposits travertine. As the travertine builds up into a large mound, known as a dam, the waterfall advances downstream. Most waterfalls erode their base, collapse, and migrate upstream, but travertine dams do the opposite: they deposit more and more new rock, building downstream. Look for the spongy-looking, porous, chalk-colored travertine on the cliff face and downstream for hundreds of feet below the falls. You can map the former positions of the stream channel by mapping the location of the travertine. This process of travertine deposition is ongoing; the

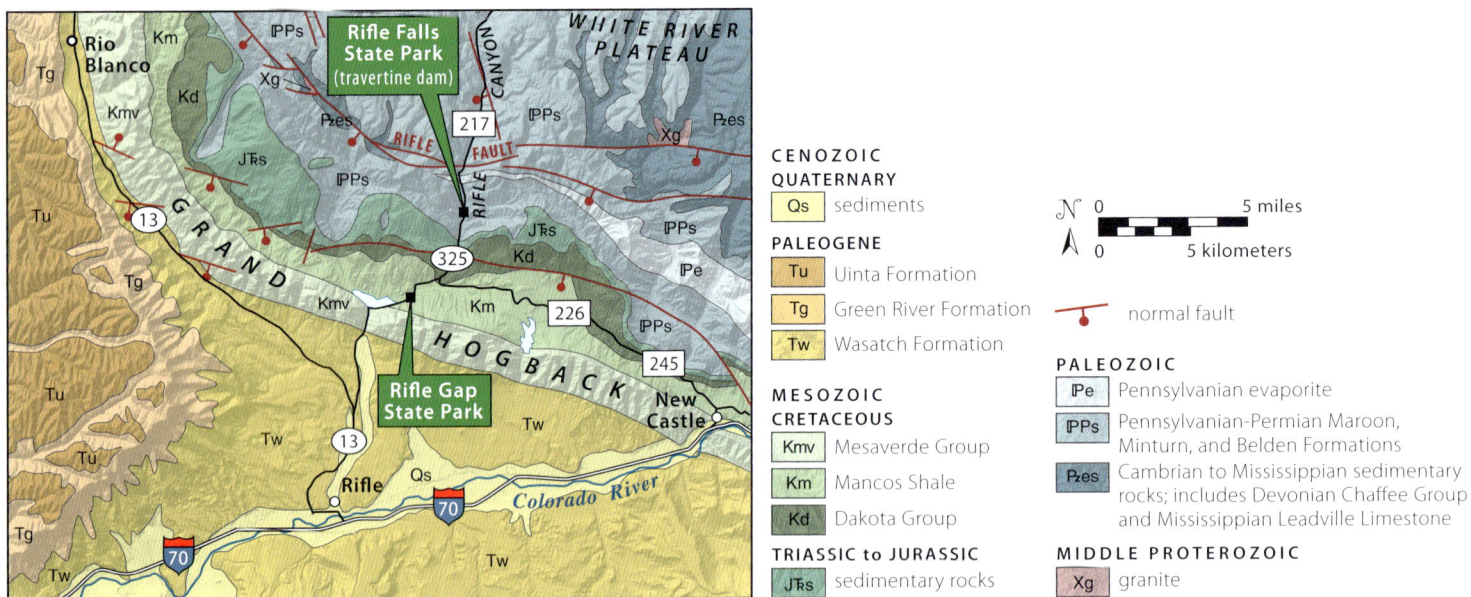

CENOZOIC
QUATERNARY
- Qs sediments

PALEOGENE
- Tu Uinta Formation
- Tg Green River Formation
- Tw Wasatch Formation

MESOZOIC
CRETACEOUS
- Kmv Mesaverde Group
- Km Mancos Shale
- Kd Dakota Group

TRIASSIC to JURASSIC
- JℝRs sedimentary rocks

PALEOZOIC
- ℙe Pennsylvanian evaporite
- ℙPs Pennsylvanian-Permian Maroon, Minturn, and Belden Formations
- ℙzes Cambrian to Mississippian sedimentary rocks; includes Devonian Chaffee Group and Mississippian Leadville Limestone

MIDDLE PROTEROZOIC
- Xg granite

normal fault

From exit 90 on I-70 at Rifle, head north on CO 13 for 4 miles and then turn east (right) on CO 325. You will pass through a gap in the Grand Hogback and then continue another 6 miles to Rifle Falls State Park, which requires a pass for day use. The park can be very busy during summer, so plan accordingly. Beyond the park, a gravel road follows the creek upstream into the canyon at Rifle Mountain Park, where towering overhung cliffs of limestone challenge even the best rock climbers.

East Rifle Creek cascades over the lip of travertine at Rifle Falls. —Marli B. Miller photo

Close-up view of the travertine at Rifle Falls.
—Marli B. Miller

travertine deposited today will be built upon in the future to the marvel of curious visitors for many years to come.

Beneath the falls are several large caves, including one that is up to 90 feet in length and requires a flashlight to explore. These caves form when water, percolating through the porous travertine from above, dissolves some of the carbonate material, transporting it downstream.

On the way to Rifle Falls State Park, you drive through the Grand Hogback, a ridge of tilted layers of the Mesaverde Group, which was deposited along the edge of the Western Interior Seaway as it was retreating in Cretaceous time. The Grand Hogback rims the western edge of the Rockies in the same way that the Dakota Hogback lines the Front Range on the east side. Both hogbacks formed as the core of the mountains rose during the Laramide orogeny.

Inside an entrance of a cave in the travertine at Rifle Falls. Note how the walls are decorated by stalactites and flowstone. —Marli B. Miller photo

The West Elk Breccia of the Dillon Pinnacles rises above Blue Mesa Reservoir. The yellow-tan layers on the left-hand cliff face are the underlying Cretaceous sedimentary rock. The Blue Mesa Tuff is the darker layer above the pinnacles. —Magdalena S. Donahue photo

28 Dillon Pinnacles in Curecanti National Recreation Area
Ashy Spires above Blue Mesa Reservoir

The Dillon Pinnacles rise about 600 feet above Blue Mesa Reservoir on the Gunnison River. The pinnacles are formed of West Elk Breccia, a jumbled mix of variously sized, angular volcanic fragments held within a matrix of fine volcanic ash and mud. Sometime between 34 and 29 million years ago, a powerful, destructive lahar—a volcanic mudflow—swept from its volcanic vent in the West Elk volcanic field. The mudflow incorporated rock from the vent and nearby country rock into one tumbled mass that became the breccia.

This region has a long volcanic history with many volcanic deposits. Here at the northern edge of the San Juan Mountains, volcanic rocks from the San Juan volcanic field to the south and West Elk volcanic field to the northeast are interlayered. The Blue Mesa Tuff, erupted from the San Juans, flowed over the top of the West Elk Breccia about 28 million years ago.

When the majority of San Juan volcanic activity ceased around 26 million years ago, the landscape began to erode. The Gunnison River cut down into bedrock, exposing more and more of the volcanic breccia. The fine, ashy matrix of the West Elk Breccia can easily erode, but the larger clasts of rock mixed within are more resistant. These resistant cobbles and boulders offer shelter to the underlying soft, ashy layers, acting as umbrellas to the columns of rock. As this process continues, tall spires of rock emerge, all topped with their resistant caprock. Underlying these spectacular volcanic rocks are Cretaceous sedimentary rocks.

Large feldspar crystals formed when the magma was deep underground, and then the rest of the magma crystallized rapidly when it intruded near the surface. —Magdalena S. Donahue photo

29

Crested Butte
A Laccolith Stands Alone

As you approach the Central Rockies from any direction, the mountains seem to build on themselves, getting bigger, taller, and more massive as you drive into them. Crested Butte, on the other hand, stands alone, rising high above the sagebrush-covered terraces of glacial outwash that line the East River. Valleys separate Crested Butte from other mountains on all but its northwest side, where a high-but-still-distinct pass provides access to Gothic and the Rocky Mountain Biological Laboratory. Crested Butte's isolation is no accident, no whim of the forces of erosion.

Crested Butte, as well as the nearby lonely peaks of Mt. Whetstone and Gothic Mountain, is a laccolith, an igneous intrusion. When magma rises into shallow regions of the crust and then intrudes parallel to bedding layers, it bows up overlying rock layers to make a mushroom-shaped igneous body. The overlying sedimentary rocks have since eroded away, leaving behind the more resistant igneous rock. Crested Butte's quartz monzonite and granodiorite contain large crystals of feldspar that formed early, before the magma injected up into the shallow rock layers. These large crystals are surrounded by a finer-grained mass that

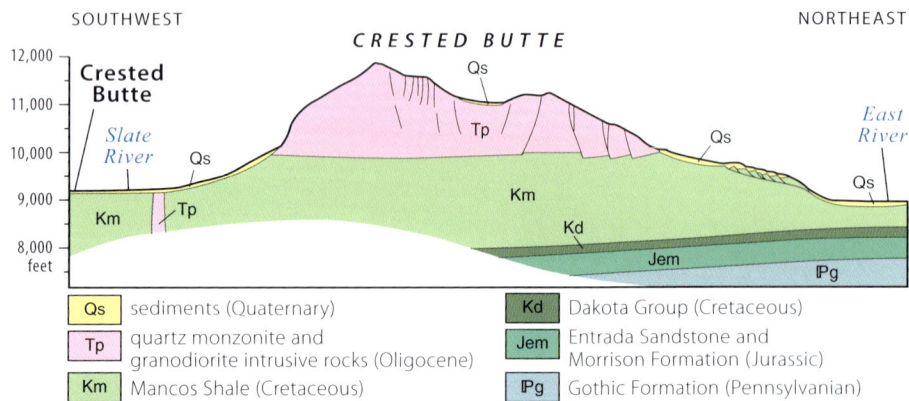

Crested Butte is a laccolith, an igneous intrusion that spread laterally, parallel to the bedding in the Cretaceous sedimentary rocks.
—Gaskill and others, 1991

Qs sediments (Quaternary)	**Kd** Dakota Group (Cretaceous)
Tp quartz monzonite and granodiorite intrusive rocks (Oligocene)	**Jem** Entrada Sandstone and Morrison Formation (Jurassic)
Km Mancos Shale (Cretaceous)	**Pg** Gothic Formation (Pennsylvanian)

70

cooled quickly near the surface. Although more resistant than the underlying Mancos Shale, the butte is riddled with vertical cracks that cause instability. It was further eroded by glaciers during the last ice age.

The nearby town of Crested Butte supplied local silver and coal mines beginning in the 1870s. A century later, molybdenum was the metal of interest. Metamorphism and hydrothermal alteration of the sedimentary rocks during intrusion of the Crested Butte laccolith in the Oligocene Epoch formed of a zone of molybdenum ore at Mt. Emmons, west of Crested Butte. The mineralization occurs in the Mancos Shale and Mesaverde Group.

The steep western face of Crested Butte demonstrates the rapid weathering of the igneous rock. —Magdalena S. Donahue photo

SAN JUAN MOUNTAINS AND RIO GRANDE RIFT

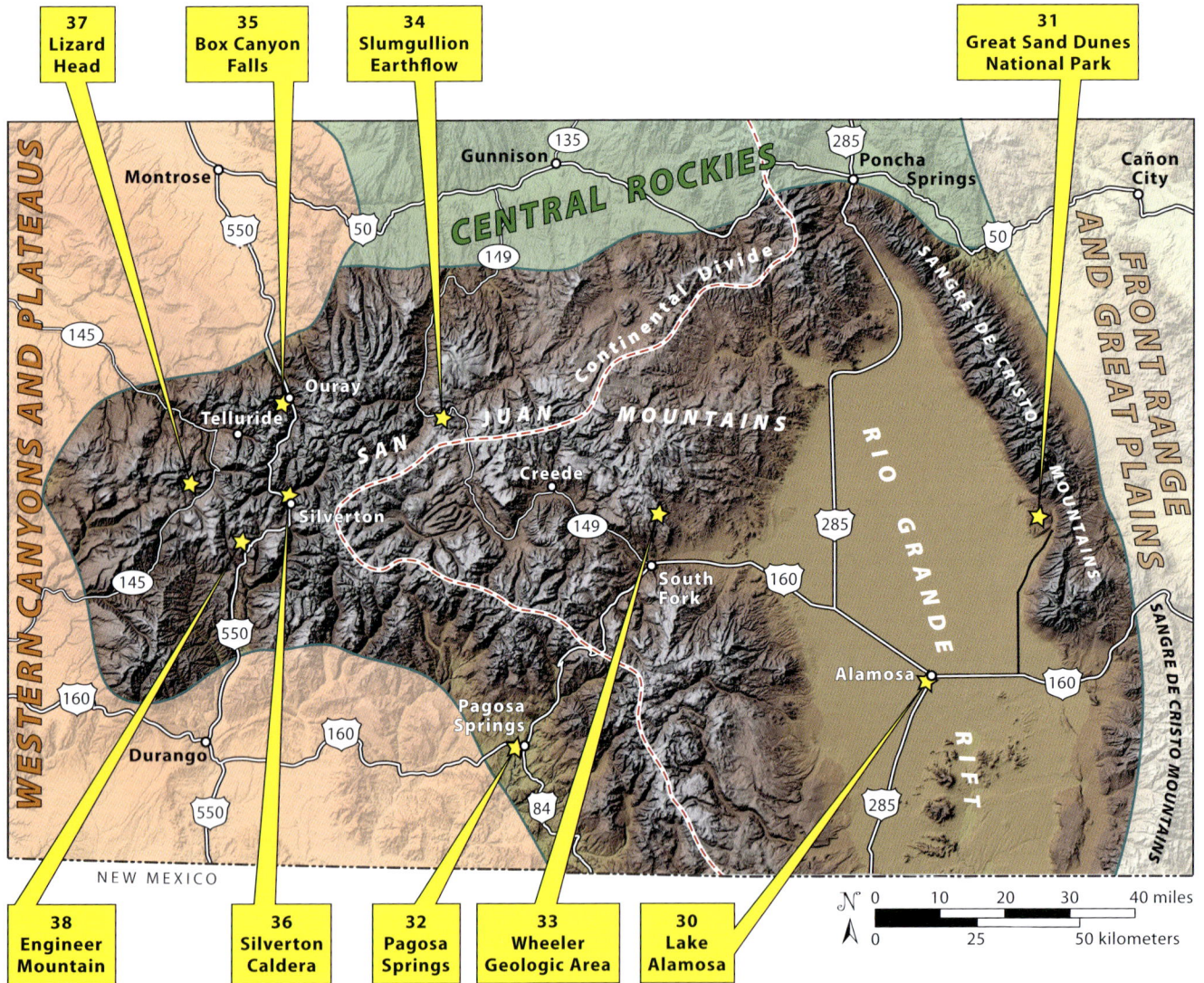

37
Lizard Head

35
Box Canyon Falls

34
Slumgullion Earthflow

31
Great Sand Dunes National Park

38
Engineer Mountain

36
Silverton Caldera

32
Pagosa Springs

33
Wheeler Geologic Area

30
Lake Alamosa

WESTERN CANYONS AND PLATEAUS

CENTRAL ROCKIES

FRONT RANGE AND GREAT PLAINS

SAN JUAN MOUNTAINS

SANGRE DE CRISTO MOUNTAINS

SANGRE DE CRISTO MOUNTAINS

RIO GRANDE RIFT

Continental Divide

Montrose

Gunnison

Poncha Springs

Cañon City

Ouray

Telluride

Creede

Silverton

South Fork

Pagosa Springs

Durango

Alamosa

NEW MEXICO

135
285
50
50
149
550
145
145
550
160
160
149
285
160
160
84
550
285

N

0   10   20   30   40 miles
0   25   50 kilometers

72

The impressive volcanic edifices of the San Juan Mountains dominate the high topography of southwest Colorado. The mountains are a complex of calderas of Oligocene age built upon a foundation of some of the oldest rocks in North America. Between 38 and 23 million years ago, more than twenty volcanoes and calderas erupted huge volumes of volcanic materials, including ash, lava, breccia, and tuff. Fluids from the hot magmas deposited some of Colorado's rich ores, most notably at Silverton.

The Rio Grande rift, a series of interconnected basins that opened in Oligocene to Miocene time, slashes through the Rocky Mountains, on the east side of the San Juans. The North American continent began to split apart along the rift, overlapping in time with the San Juan volcanism. The split has not yet been successful, but the chain of linear valleys through Colorado and New Mexico reminds us of its intentions.

The Continental Divide snakes through the San Juan Mountains. The west side of the mountains drains into the San Juan, Animas, and Dolores Rivers, flowing eventually into the Colorado River and west to the Pacific Ocean. The east side forms the uppermost headwaters of the Rio Grande, one of the largest rivers in North America. It captures the majority of the southern Rocky Mountain water, carrying it to the Gulf of Mexico.

A string of basins formed along the Rio Grande Rift.

Lake Alamosa was the latest in a series of lakes that occupied the San Juan Valley in Pliocene to Pleistocene time. —Modified from Machette 2007

30 San Luis Valley
Flat Floor of Lake Alamosa

The San Luis Valley, an oblong high-elevation plain between the Sangre de Cristo Mountains to the east and the San Juan Mountains to the west, is surprisingly flat. The valley is part of the Rio Grande rift, a series of north-trending, fault-bound basins between Colorado and Texas that began forming 30 million years ago when the crust of North America was stretched in an east-west direction. These rift basins are up to about 10,000 feet deep, filled with layer after layer of sediment that has become loosely solidified over millions of years. These basins were at times dammed and filled with lakes, with Lake Alamosa being the last lake to occupy San Luis Valley. The extreme flatness of the San Juan Valley is the most visible evidence of the lake, but spits and sand bars exist on the sides of the valley, and lagoon deposits are present on the northern margin of the San Luis Hills. In addition, drill cores into the valley bottom reveal that loose sediments eroded from the mountains are interbedded with lake deposits.

The San Luis Valley is also home to the upper reaches of the Rio Grande, the river that stretches nearly 2,000 miles from the high peaks of the San Juans through New Mexico to Texas, where it forms the US–Mexico border before draining into the Gulf of Mexico. About 5 to 4 million years ago, the southern end of the San Luis Valley was blocked by basalt flows originating from volcanic vents along the Colorado–New Mexico border and flowing north around the edges of the Oligocene-age volcanic San Luis Hills. Because the valley could no longer drain to the south, it was filled with a large lake.

Lake Alamosa is postulated to have been one of the world's largest high-elevation lakes (Alamosa sits at 7,540 feet elevation). The last time the lake reached its maximum volume, about 450,000 years ago, it extended nearly 65 miles north to south and was 30 miles wide. As the lake continued to fill with water due to the wet glacial climate, the lake eventually overflowed through volcanic edifices

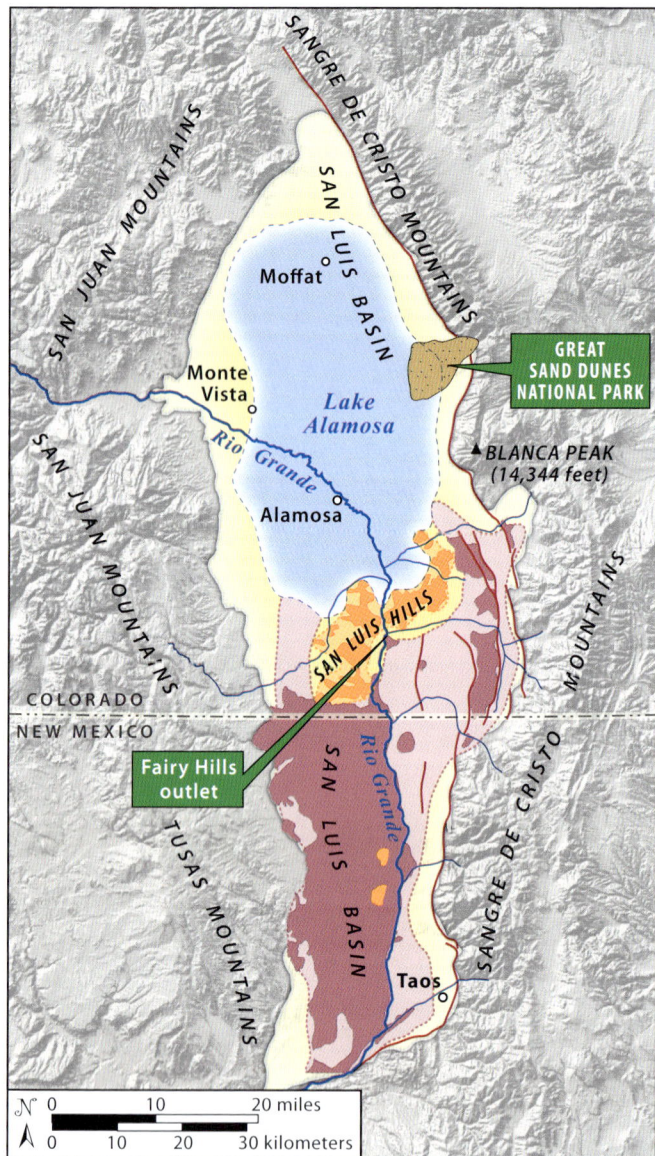

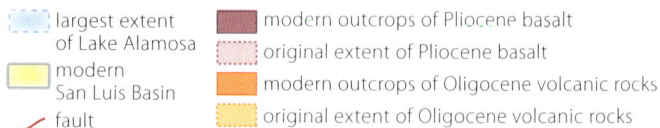

The Sangre de Cristo Mountains rise high above sandhill cranes feeding on the flat floor of the San Luis Valley at Monte Vista National Wildlife Refuge. —Leith Edgar photo, US Fish and Wildlife Service

called the Fairy Hills, about 15 miles north of the Colorado–New Mexico border. The lake completely drained by 440,000 years ago. The overflow eroded a deep channel that persists to this day and hosts the Rio Grande.

Despite its high elevation, the flat valley floor supports agriculture. Farmers irrigate crops with groundwater pumped from aquifers in the valley sediments. Water infiltrates the loose valley sediments until it reaches a layer of shale that forms an impermeable barrier. The shale was once lake mud deposited in deep areas of a lake. The water accumulates in the overlying porous and permeable sediment, becoming what is called a perched aquifer. Finding these aquifers can be hit or miss, given that the environments in which the sediments were deposited were constantly shifting over time.

The Great Sand Dunes rise above the flat valley floor, trapped in front of the southern Sangre de Cristo Mountains.
—Magdalena S. Donahue photo

31 Great Sand Dunes National Park and Preserve
Anchored by Water

The tallest sand dunes in North America rise 750 feet above the floor of the San Luis Valley. The dunes, protected as Great Sand Dunes National Park and Preserve, are located on the eastern edge of the San Luis Valley, nestled in a bend beneath the precipitous Sangre de Cristo Mountains. The dunes and a surrounding sand sheet cover about 30 square miles and are estimated to be composed of more than 5 billion cubic meters of sand. Where does all the sand come from and why doesn't the wind blow it away?

After Lake Alamosa (site 30) drained about 440,000 years ago, winds from the southwest blew across the drying lakebed, carrying sand to the northeast, where it got trapped at the base of the Sangre de Cristo Mountains. Exactly how long it took the huge pile of sand to accumulate is unknown. Erosion of two mountain ranges, the San Juans to the west and the Sangre de Cristos to the east, contributed to the sediment. Grass and shrubs grow on the sand sheet that extends out from the main dune field on three sides. Although not as remarkable as the towering dunes, the sand sheet contains nearly 90 percent of the sand in the park and is the primary source of sand for the dunes.

During winter and spring, and after monsoonal storms in summer, streams and standing water can be found in the dune field. Digging a few inches into the dunes reveals wet sand year-round. Wet sand doesn't blow away like dry sand, so this wet interior helps anchor the dunes in place. The water also allows grasses and shrubs to take root at the edges of the dune field, reducing the ability of the wind to move the sand. Finally, as the water at the surface evaporates, minerals precipitate to form a hard crust. Look for the whitish crust at the side of the road as you approach the park. The crust forms anew with each drying season.

Although the predominant wind direction is from the southwest, winds from the east occur during storms, and together, these opposite-direction wind patterns cause the dunes to grow vertically as their sand material gets blown back and forth, piling up on itself. Though generally quite stable, some dunes are migratory, and many change shape seasonally.

Ripples in the sand at Great Sand Dunes National Park. —Marli B. Miller photo

Medano Creek flows from the Sangre de Cristo Mountains in the distance. —Meg Miller photo

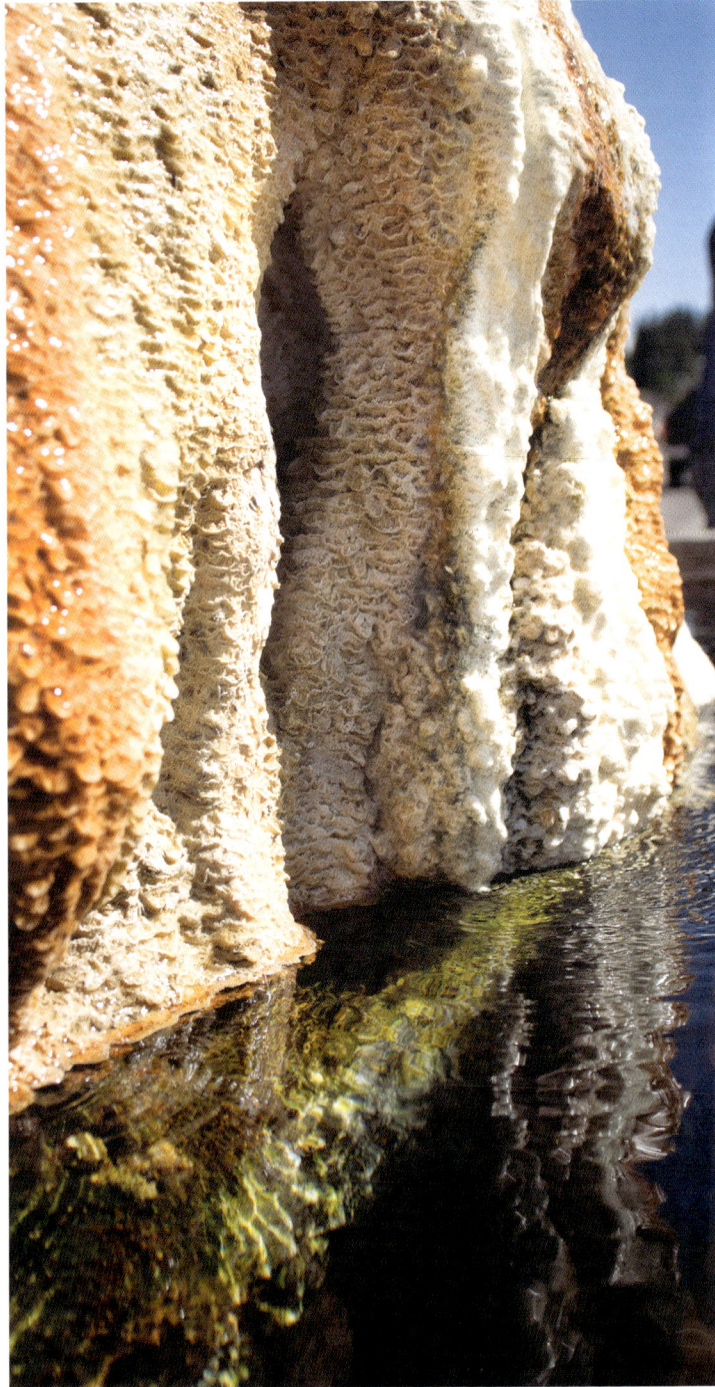

32 Pagosa Springs
Colorful Towers of Siliceous Sinter

The town of Pagosa Springs derives its name from the Ute word *pagosah*, meaning "healing" or "boiling waters." Pagosa's main hot spring, on the banks of the San Juan River, discharges about 700 gallons per minute. The spring water contains more than just H_2O. Hydrogen sulfide, known for its rotten-egg smell, is easily noticed. Other dissolved solids carried by hot waters include silica (quartz), sodium, chloride, and potassium. When the hot, mineral-saturated waters cool and become oversaturated, minerals are precipitated. At Pagosa Springs, precipitation forms strange towering or draping deposits of siliceous sinter, a type of opal sometimes known as geyserite. Look for these in the public springs near the San Juan River in downtown Pagosa Springs, as well as in the private developed pools. The deposits of geyserite are so voluminous they have forced the San Juan River to curve around the springs. The bend in the river is visible from the municipal parking lot in downtown Pagosa Springs.

The spring water reaches the surface at temperatures that can exceed 150 degrees Fahrenheit, providing a geothermal heat source for the town. The heat, which comes from relatively shallow hot mantle rock, has persisted since the Oligocene volcanism that created the San Juan volcanic field, visible on the sweeping skyline north of town. The groundwater absorbs heat from the rocks as it flows through faults and fractures created during the region's active volcanic history. The heated water then rises, mixing with the melting snow and rainwater percolating down through the ground. The groundwater is stored in a Dakota Group sandstone aquifer, which is overlain and confined by the impermeable Mancos Shale. Both of these units are present at the surface around Pagosa Springs, and the springs emerge from fractures in the Mancos Shale.

Fragile and constantly changing ripples on the hot spring deposits form where water has flowed or dripped down the side of the spring. —Magdalena S. Donahue photo

A spout of hot water, artificially piped vertically, fountains from a hot spring in downtown Pagosa Springs. Multicolored algae and bacteria thrive in the warm, mineral-rich waters, growing into scummy mats and slimes that coat the sides of many hot springs. —Magdalena S. Donahue photo

La Garita caldera boundary

Tfc approximate extent of Fish Canyon Tuff

N 0 10 20 30 40 50 miles
0 25 50 kilometers

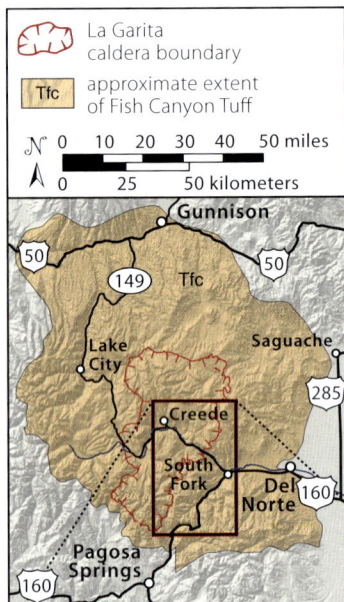

You can see outcrops of Fish Canyon Tuff along US 160 south of South Fork. The tuff erupted from La Garita caldera.

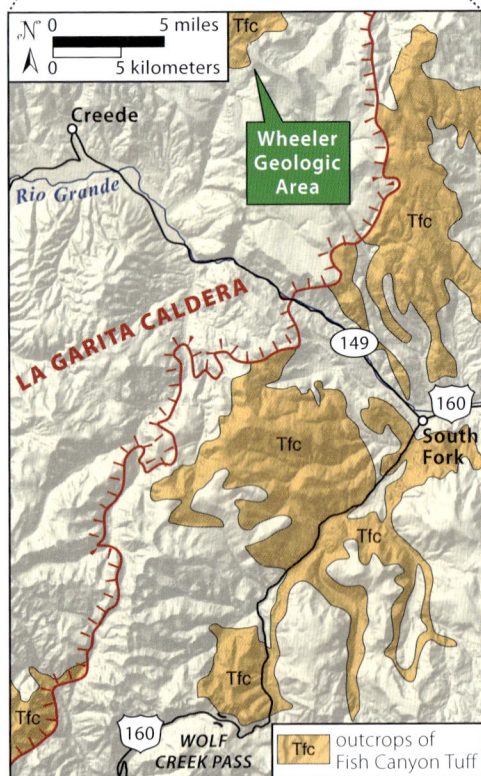

N 0 5 miles
0 5 kilometers

Tfc outcrops of Fish Canyon Tuff

33 Wheeler Geologic Area
Fish Canyon Tuff of La Garita Caldera

More than twenty calderas erupted during the Oligocene Epoch from the San Juan volcanic field in southwestern Colorado. The 27.8-million-year-old La Garita caldera is the largest of them and possibly the largest caldera eruption on Earth. The La Garita event spewed out as much as 3,000 cubic miles of magma in violent pyroclastic flows—mixtures of hot ashes, volcanic fragments, and gases—that poured down the caldera flanks. Some of these flows cooled to form the voluminous Fish Canyon Tuff. The caldera collapsed after this eruption, and smaller volcanic flows erupted from vents nested within the collapsed area over the next million years as the underlying magma chamber cyclically refilled and erupted. The caldera is approximately 22 by 50 miles, extending from Saguache Park, across the town of Creede, and south into the Weminuche Wilderness west of Wolf Creek Pass.

Two places provide outstanding views of the caldera's rocks: Wolf Creek Pass and the Wheeler Geologic Area. At Wolf Creek Pass, where US 160 crosses the Continental Divide, you can see roadside exposures of the white-buff Fish Canyon Tuff. You can also see the tuff north of Wolf Creek Pass along US 160 just south of the town of South Fork.

Named for Captain George Wheeler, leader of an early western geologic survey team, the Wheeler Geologic Area was Colorado's first national monument in 1908. Given Colorado's amazing natural wonders, it may seem odd that this out-of-the-way site was recognized so early. When you see it, you'll realize why: it's like no place you've ever seen. Colorful striking spires of the Fish Canyon Tuff are exposed at nearly 12,000 feet elevation. The monument was never developed, however, due to its remoteness and lack of funds, and eventually the area was transferred to the La Garita Wilderness. It is extremely difficult to reach the Wheeler Geologic Area; please consult with the US Forest Service for information on how to access it.

The presence of a resistant caprock shields the underlying, more erodible volcanic material of the spires. As erosion continues, the material beneath the caprock remains in place while surrounding material is removed, and eventually the spires become unstable and topple.

Fish Canyon Tuff on the west side of US 160 about 1.5 miles south of the CO 149 intersection in the town of South Fork.
—Marli B. Miller photo

Many layers of tuff and volcanic material erupted from the La Garita caldera have been carved into spectacular pinnacles at Wheeler Geologic Area. Each different color represents a unique eruption. The softer pale white and tan layers in the foreground erode more easily than the more densely welded tuffs that form the skyline.
—John Fowler photo, Creative Commons license 2.0

81

Slumgullion Earthflow
A Slow-Moving Landslide Lubricated by Snowmelt

The Slumgullion earthflow, 4 miles south of Lake City, is a type of landslide in which slowly moving sediment is pulled downward by gravity. The earthflow, which is approximately 4.2 miles long, 0.25 mile wide, and about 400 feet deep, is still moving, and scientific measurements show it travels approximately 23 feet per year. Scientists determined the earthflow began moving between 1,000 and 1,600 years ago by measuring the amount and thickness of soil developed on the upper, oldest parts of the landslide. About 700 years ago, the landslide dammed the Lake Fork of the Gunnison River, forming Lake San Cristobal, Colorado's second-largest naturally occurring lake. The age of the slide was determined from radiocarbon dating of wood fragments from the base of the landslide deposits and soil humus from beneath the landslide deposits. The toe that dams the lake is now inactive, and the Lake Fork has incised around the slide where it abuts the opposite mountainside.

The flowing slide material includes volcanic basalt, rhyolite, and andesite, erupted from a caldera in the San Juan volcanic field in Oligocene time. The bright colors of the volcanic rock are a result of hydrothermal alteration, a process in which superheated waters permeated these units and changed the minerals. The hot fluids precipitate interesting minerals such as pyrite (fool's gold) and chalcedony,

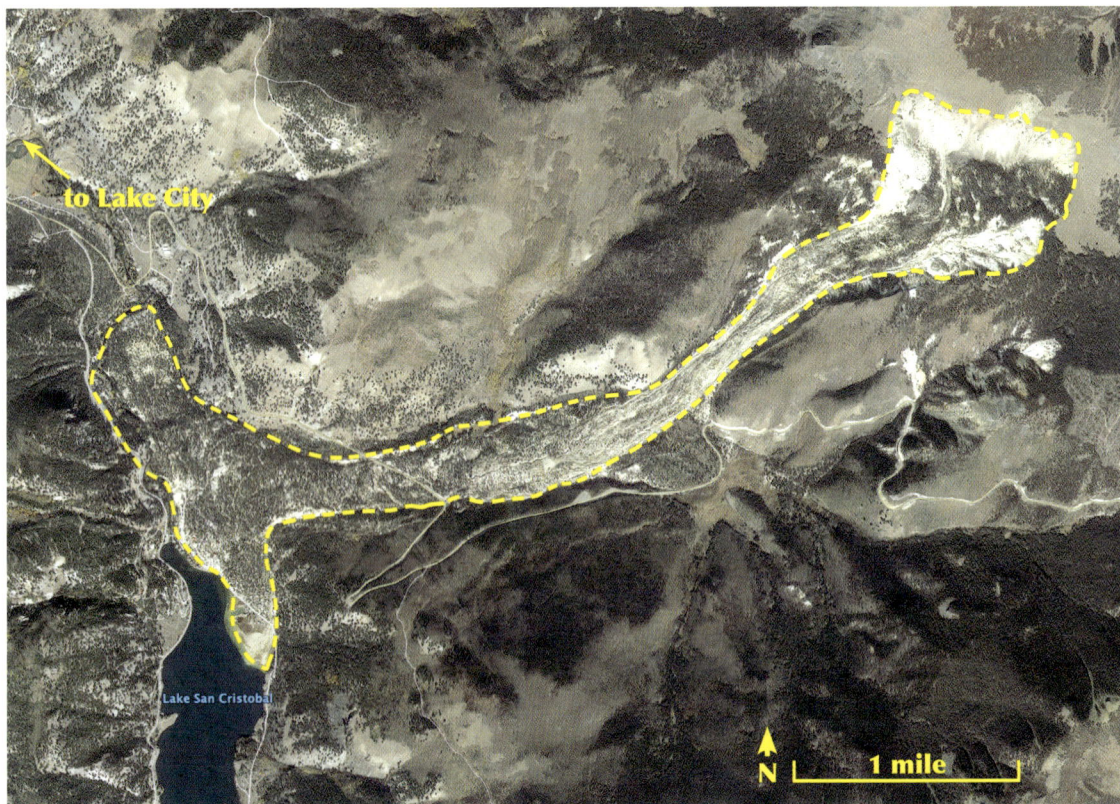

The Slumgullion earthflow is outlined in yellow on this satellite view. The flow is moving from its cliffy scarp in the right (northeast) section of the photo, oozing southwest in a linear path down to the valley floor, where it spreads out into a broad, spatula-shaped toe. These deposits have dammed the Lake Fork of the Gunnison River, forming Lake San Cristobal, visible in the lower left (southwest) corner of the image.
—Base satellite image from Google Earth

The upper scarp of the Slumgullion earthflow is visible above the forested hillslopes. This raw, unvegetated cliff face is constantly caving, and rock and earth move downward with the rest of the main flow—the entire lumpy foreground of the photo. —Magdalena S. Donahue photo

a form of quartz with submicroscopic crystals. Alteration also formed smectite clays, jarosites (hydrous sulfate minerals), and gypsum, all of which are very mechanically weak, thus enhancing the landsliding.

The younger upper portion of the earthflow is actively thickening and spreading transversely. Though movement is not visible to the human eye, you can see tilted, disturbed trees and abrupt scarps on the hillslope. The most rapid movement occurs each spring when snowmelt lubricates the debris. Tension cracks and thrust faults within the slide change how water flows, and ponds sometimes form in the jumbled material.

Waterfall in Box Canyon as seen from the metal stairway. The canyon walls consist of vertical quartzite of the Uncompahgre Formation.
—Marli B. Miller photo

35 Box Canyon Falls near Ouray
The Great Unconformity Gets Boxed In

The 500-foot-long Falls Trail on the southwest side of Ouray enters a 20-foot-wide canyon that ends at Box Canyon Falls, where Canyon Creek spills over the Uncompahgre Formation. The east-west-trending Ouray fault crosses the canyon, placing Paleozoic rocks on the north side of the canyon in contact with quartzite of the Proterozoic-age Uncompahgre Formation on the south side. Several hundred yards of movement occurred on this fault to bring these two very different-aged rock formations together. As you walk southward into the canyon, you cross the fault and enter the steeply dipping Uncompahgre Formation, which was originally deposited in a marine basin 1.7 to 1.4 billion years ago. A perched metal catwalk and stairway make for an exhilarating walk to the creek bottom, just below the 285-foot waterfall. The best, albeit limited, view of the falls is from the stairway. Look for black swifts, which nest in the steep canyon walls.

If you hike the steep but short quarter-mile-long High Bridge Trail, you get a bird's-eye view of the canyon, along with a fantastic display of the Great Unconformity. At the bridge, the vertical Uncompahgre Formation is overlain by the gently tilted Elbert Formation of Devonian age. Above the Elbert are several carbonate units, the Ouray and Leadville Limestones of Mississippian age. This exposure is significant because it indicates that the Cambrian and Ordovician rocks, which we would otherwise expect to overlie the unconformity, were eroded away before deposition of the Elbert Formation in the Devonian Period.

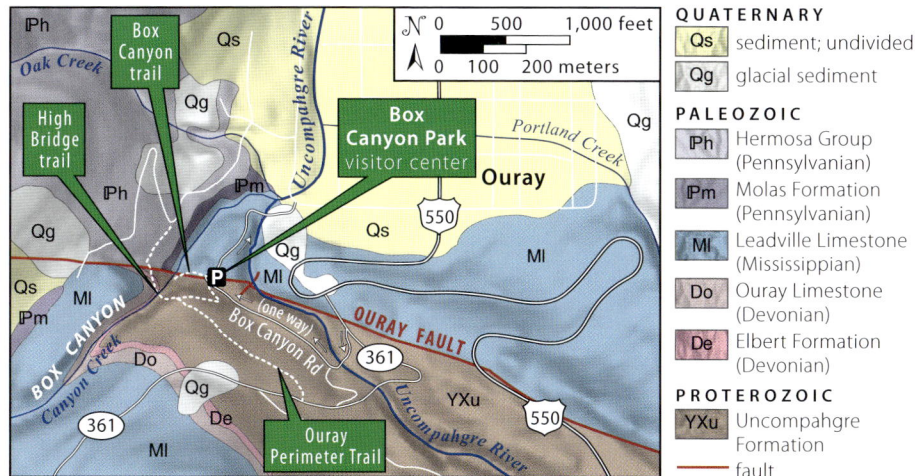

To reach Box Canyon Park, head south from Ouray on US 550, the Million Dollar Highway. Just past the first switchback, turn right on Camp Bird Road (County Road 361), and then immediately right again on Box Canyon Road, a one-way road. The park has a small fee and doesn't allow dogs. The town of Ouray hosts natural hot springs, with heated waters rising through the Ouray fault. These hot waters are piped into the Ouray Hot Springs Park.

QUATERNARY

Qs	sediment; undivided
Qg	glacial sediment

PALEOZOIC

ℙh	Hermosa Group (Pennsylvanian)
ℙm	Molas Formation (Pennsylvanian)
Ml	Leadville Limestone (Mississippian)
Do	Ouray Limestone (Devonian)
De	Elbert Formation (Devonian)

PROTEROZOIC

YXu	Uncompahgre Formation
	fault

The Great Unconformity, here with Devonian sedimentary rocks on top of nearly vertical Proterozoic rock, as seen from the High Bridge in Box Canyon. —Marli B. Miller photo

36

Silverton Caldera and Silverton Mining Heritage Center
Precious Metals in Volcanic Rock

Silverton, located in the Animas mining district, is one of the great precious metal areas of Colorado. Major mining operations here lasted from 1860 to 1960 and yielded more than 1,655,000 ounces of gold. The Animas and nearby Eureka regions also produce silver, lead, copper, and zinc. The metals occur in volcanic rocks erupted from the Silverton caldera of the San Juan volcanic field.

In Eocene time, before the volcanic field existed, the land surface bulged upward as magma chambers filled. Erosion of the newly elevated land led to the deposition of the Telluride Conglomerate. Boulders more than 3 feet in diameter rolled into place as the rapidly uplifted region eroded. Volcanic rocks erupted over the conglomerate beginning about 38 million years ago. The change from the conglomerate to the overlying volcanics records the transition from an eroding landscape, where rock material was being eroded and washed away, to a constructive landscape, where material was being added to a landscape.

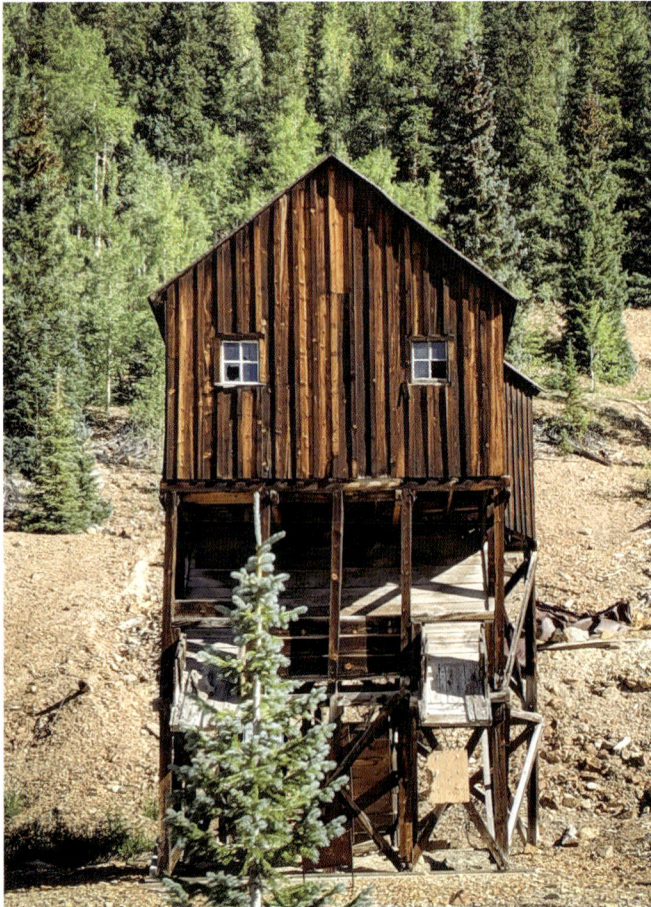

The San Juan Mountains are littered with standing and degrading mining structures like this one near Silverton, showing the incredible engineering that was required to grow and sustain mining in such rugged topography.
—Magdalena S. Donahue photo

Closeup of Telluride Conglomerate. —Magdalena S. Donahue photo

Mining area on Kendall Mountain, just south of Silverton, showing chemically altered volcanic rocks. —Marli B. Miller photo

The Silverton caldera erupted about 27 million years ago. Calderas form during massive and catastrophic events. When enough of the magma has erupted, the overlying lid of already-erupted volcanic rock collapses into the now empty magma chamber, forming a caldera. The volcano continues to vent through the fractures that circle the collapsed center.

The Silverton caldera walls include altered lavas and dacite to rhyolite tuffs collectively called the San Juan Formation. Hydrothermal fluids circulating through the volcanic rocks in the ensuing millions of years deposited minerals such as chlorite, epidote, and pyrite as well as the more precious metal ores. The alteration of volcanic material produced bright reds, oranges, and yellows, clearly visible on steep cliffs as you drive into the collapsed caldera center in which the town of Silverton nestles. Red Mountain Pass, on US 550 north of Silverton, takes its name from the color of these altered rocks. The walls and peaks to the east, north, and west of Silverton are composed of the San Juan Formation. To the south lies the Sultan Mountain stock, a greenish granitic intrusive associated with the Silverton caldera.

The San Juan County Historical Society Mining Heritage Center operates out of a historic jailhouse in Silverton. The mining museum displays local and regional minerals and ore specimens and historic photographs from the region's many mining operations. It also displays the early mining and surveying tools and methodologies, and many artifacts used in mining in the San Juan Mountains since the 1800s.

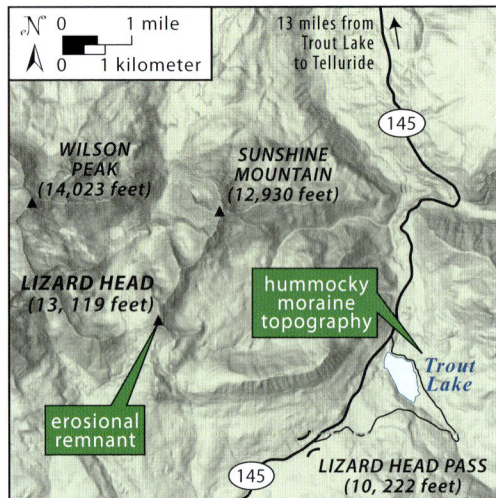

Lizard Head is visible from an overlook at Lizard Head Pass on CO 145.

37

Lizard Head
A Survivor (Just Barely) of Pleistocene Erosion

Lizard Head, one of the most striking peaks in southwest Colorado, rises starkly from the valley of Lizard Head Pass. You can see the narrow, 13,119-foot spire to the north from CO 145, the highway that traverses the pass. Lizard Head is one of the hardest, most dangerous peaks to climb in Colorado because the pinnacle is constantly eroding, breaking, and falling into talus, steep slopes of loose, broken rock.

Lizard Head is often mistaken as a remnant volcanic plug or neck, but it was actually formed through the erosion of two completely different layers of rock. The base of the spire is composed of Telluride Conglomerate, which is overlain by approximately 500 feet of Oligocene ash-flow tuffs of the San Juan Formation. The dark hillslopes below Lizard Head are the easily erodible Mancos Shale, a Late Cretaceous unit that was deposited as offshore muds when this region was inundated by the Western Interior Seaway. The loose, crumbly, and swelling nature of these clays render this unit highly susceptible to erosion. As Pleistocene glaciers and meltwater streams eroded the landscape, they easily ate into the shale and gradually chipped away at areas capped by harder rocks, such as the volcanic tuff.

Lizard Head is all that remains of the hard rock here. Evidence of glaciation is also visible. Moraine deposits, composed of the unsorted, unconsolidated materials that were scraped off and incorporated into the ice and then dumped when the ice melted, can be seen as the hummocky, hilly topography near Trout Lake, on the east side of Lizard Head Pass.

The spire of Lizard Head is tuff of the Oligocene-age San Juan Formation. It overlies the Eocene-age Telluride Conglomerate, mostly covered by talus and glacial deposits here. The black slopes at the base of the mountain are Mancos Shale, covered in places by glacial deposits. Hummocky glacial topography is visible at the lower left. —Magdalena S. Donahue photo

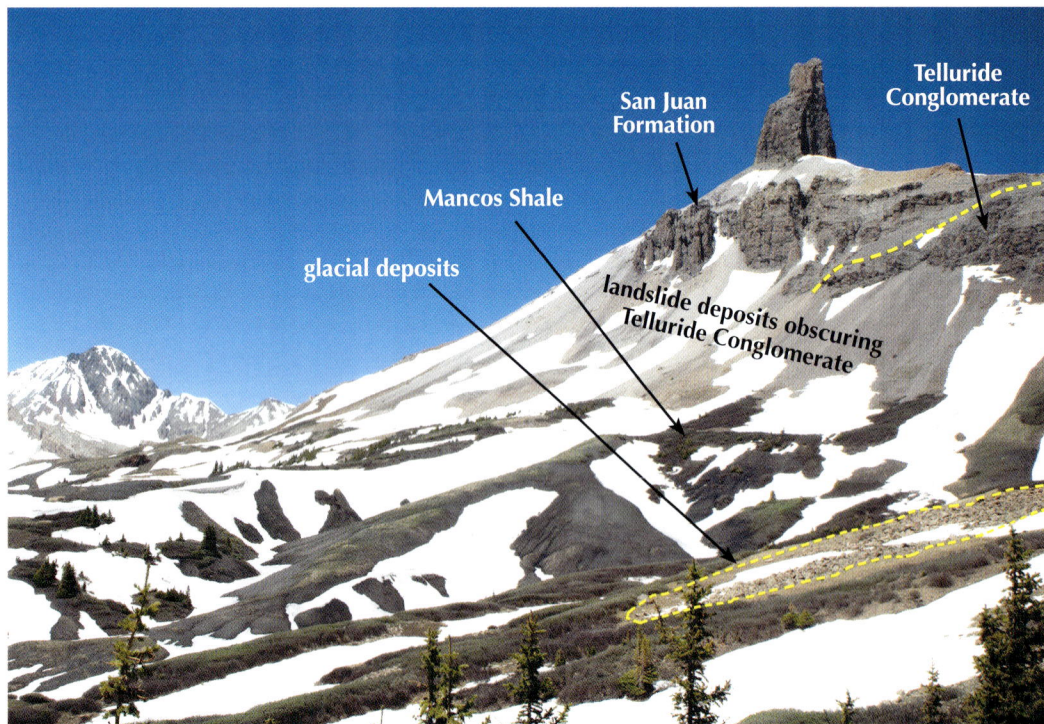

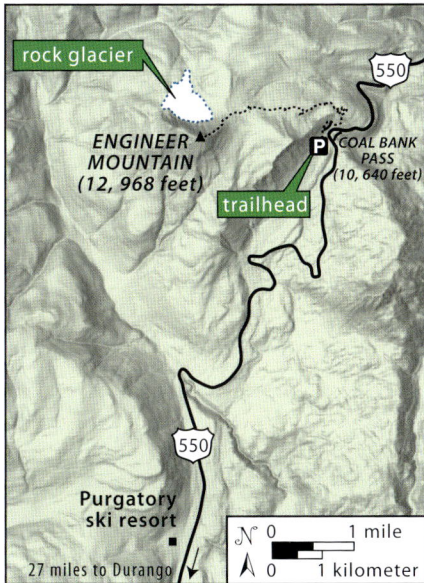

The trail to the summit of Engineer Mountain begins at Coal Bank Pass.

38 Engineer Mountain
A Sill Exposed by Ice

Engineer Mountain, a 12,968-foot peak rising just north of the Purgatory/Durango Mountain ski resort, extended above the ice during the Pleistocene ice ages. The entire Animas River valley to the east was filled with approximately 2,000 feet of glacial ice, and Engineer Mountain was one of the peaks tall enough to extend above it. The grinding ice exposed a striking band of cliffs that extends horizontally across the face of Engineer Mountain. The cliff band forms the base of a thick sill, an igneous feature in which magma intrudes largely horizontally along bedding planes within or between sedimentary rock layers. The injection of these magmas pushed up overlying layers into a dome. These

Engineer Mountain, viewed from the Million Dollar Highway, was named after and by the US Army Corps of Engineers in 1873 during the Hayden survey of the region. The bright-red Cutler Group directly underlies the sill that forms the peak and includes the band of cliffs in the vertical sides of Engineer Mountain. —Marli B. Miller photo

89

overlying rocks have been eroded from the top of Engineer Mountain, but their tilted remnants can be measured on the flanks of the nearby mountains.

The Engineer Mountain sill is composed of rhyolite with large potassium feldspar crystals. New dating of these units, published in 2015 by David Gonzales of Fort Lewis College in Durango, shows that the sill intruded the Paleozoic sedimentary rocks between 17 and 14 million years ago. The Engineer Mountain sill is associated with the nearby Barlow Creek sill, which, though eroded, once extended over tens of

This close-up, viewed from the trail to the summit, highlights the columnar jointing at the base of the sill. The joints formed when the injected volcanic material cooled and contracted. The red rock beneath the sill is part of the Pennsylvanian to Permian Cutler Group. —Marli B. Miller photo

Close-up of the rhyolite that makes up the Engineer Mountain sill. Note the large feldspar crystals. Hand lens for scale. —Marli B. Miller photo

View north over the rock glacier on Engineer Mountain's north side. —Marli B. Miller photo

miles. The vertical cracks in the sill formed during cooling of the magma. Known as columnar jointing, they form perpendicular to the cooling surfaces, in this case most apparent at the bottom of the sill. This vertical jointing is a common cooling feature, also famously seen in the Giant's Causeway in Ireland and Devils Tower National Monument in Wyoming.

As a testament to its glacial past, the north side of Engineer Mountain forms a glacially eroded cliff that rises nearly vertically about 1,000 feet. Below it, a rock glacier flows northward down the valley. Rock glaciers are bodies of broken rock debris with ice cores that, like regular glaciers, flow downhill in response to gravity.

WESTERN CANYONS AND PLATEAUS

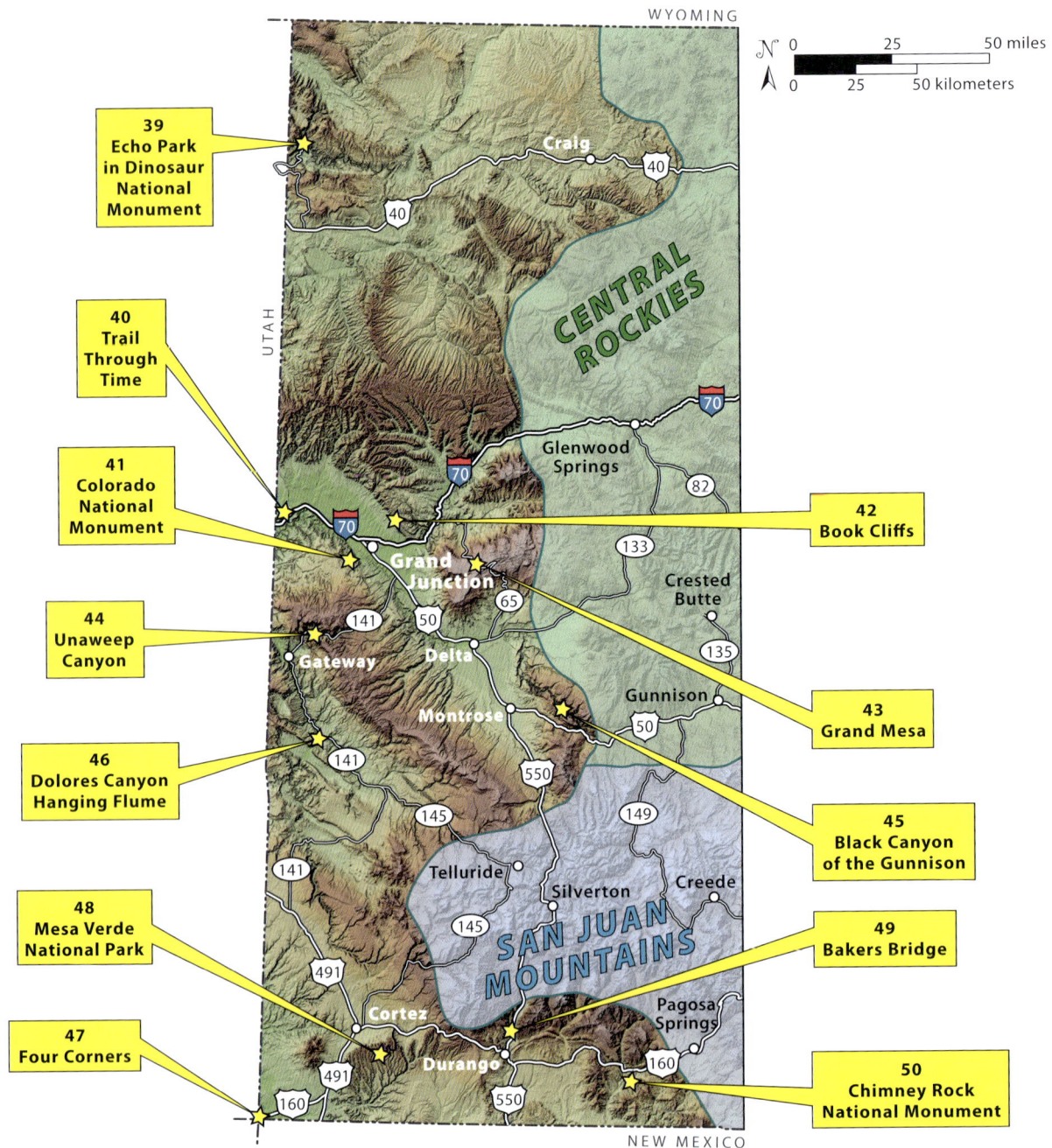

WYOMING

0   25   50 miles
0   25   50 kilometers

CENTRAL ROCKIES

UTAH

Craig

40

40

Glenwood Springs

70

82

133

Crested Butte

135

Grand Junction

65

50

141

Delta

Gunnison

50

Montrose

550

149

145

550

**39
Echo Park
in Dinosaur
National
Monument**

**40
Trail
Through
Time**

**41
Colorado
National
Monument**

**42
Book Cliffs**

**44
Unaweep
Canyon**

**43
Grand Mesa**

Gateway

**46
Dolores Canyon
Hanging Flume**

**45
Black Canyon
of the Gunnison**

SAN JUAN MOUNTAINS

Telluride

Silverton

Creede

145

**48
Mesa Verde
National Park**

**49
Bakers Bridge**

491

**47
Four Corners**

Cortez

Pagosa Springs

Durango

160

491

160

550

**50
Chimney Rock
National Monument**

NEW MEXICO

McInnis Canyons National Conservation Area along the Colorado-Utah border contains the second-largest concentration of natural arches in North America. —Bob Wick photo, Bureau of Land Management

The Colorado Plateau physiographic province extends from western Colorado into Utah, Arizona, and New Mexico. Famously hosting the many national parks of southern Utah, the Colorado Plateau is a stack of sedimentary rock layers that have remained relatively unchanged in their character and orientation since deposition in Mesozoic time. The Western Slope region of Colorado was first uplifted during the formation of the Ancestral Rockies, at which time many of the Paleozoic sedimentary layers were eroded from the basement rock.

The flat-lying rocks of Mesozoic age form many mesas in western Colorado and are an important source of petroleum, uranium, and coal. Over time, streams eroded the less-resistant shales into valleys and badlands, including adobe badlands in the Cretaceous-age Mancos Shale that do not support much vegetation. Recent uplift invigorated the region's rivers, which have incised down through the sedimentary layers, forming deep canyons.

Echo Park in Dinosaur National Monument

Confluence of the Yampa and Green Rivers

Dinosaur National Monument straddles the Colorado-Utah border. The famous quarry in the Morrison Formation that preserves more than 1,000 Jurassic-age fossils lies in the Utah portion of the monument. Visitors to Colorado's remote portion of the monument are rewarded with views of the spectacular canyons of the entrenched Yampa and Green Rivers. The incision of the rivers did not begin until around 9 to 6 million years ago, when regional uplift caused the gravity-driven water to carve downward. Once the rivers became entrenched in the resistant sandstone, there was no other way to move than to slice downward through many layers of sedimentary and volcanic rock. The Canyon of

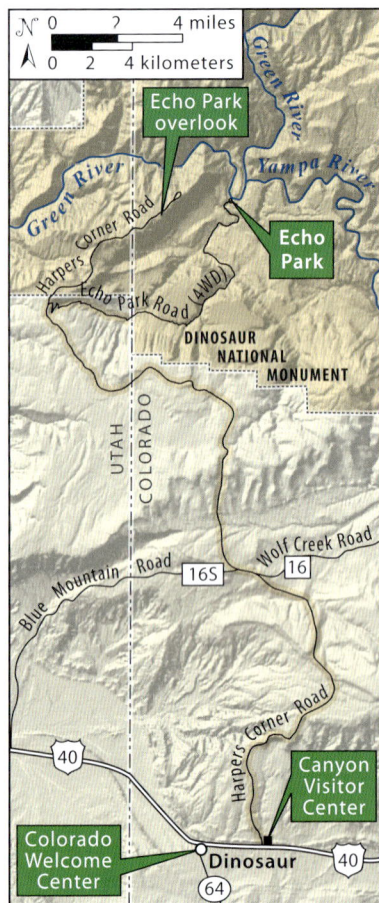

Harpers Corner Road leads 33 miles north from Dinosaur to Echo Park overlook. You can drive to the base of the canyon on Echo Park Road if the road is dry and you have a four-wheel-drive, high-clearance vehicle.

The green water of the Green River and the brown water of the Yampa, visible just to the right of the Steamboat Rock outcrop, are slow to mingle. In the distance, the deep incision of rivers into the plateau country is visible.
—Magdalena S. Donahue photo

Lodore on the Green River reaches into 766-million-year-old sandstones and shales of the Uinta Mountain Group, the oldest rock unit exposed in the monument.

Echo Park, 37 miles north of the Canyon Visitor Center near the town of Dinosaur, provides access to Steamboat Rock at the confluence of the Yampa and Green Rivers. The high walls of Steamboat Rock are the Weber Sandstone, a massive, thickly bedded, fine-grained yellowish rock that is the primary cliff-forming unit in Dinosaur National Monument. The sandstone was deposited as ancient sand dunes in middle Pennsylvanian time and is as much as 1,150 feet thick! Look for cross beds formed by winds blowing sand in the dunes.

At the confluence, the different-colored waters of the two rivers are visible as they join and mix. The Green River's water is largely blue and clear, its sediments having settled out in the lake impounded by the Flaming Gorge Dam upstream from here. The more sediment-rich waters of the Yampa tend to be brown to reddish. The rivers continue to deepen their incredible canyons with each passing year.

Steamboat Rock stands above the confluence of the Yampa and Green Rivers. —Magdalena S. Donahue photo

Along the Colorado-Utah border in McInnis Canyons National Conservation Area, the dinosaur-fossil-bearing Morrison Formation of Jurassic age is at the surface. This same formation houses the dinosaur quarry in the Utah portion of Dinosaur National Monument. McInnis Canyons is famous for the discovery of a nearly complete *Apatosaurus* in 1900. At 70 feet in length, this enormous dinosaur skeleton is displayed in the Chicago Field Museum of Natural History. The beast weighed about 30 tons when it roamed the late Jurassic floodplain. *Brachiosaurus*, *Stegosaurus*, and *Allosaurus* were also discovered in this location within the Morrison Formation.

If you are traveling along I-70 through this vast, lonely region, stop at the Trail Through Time, a 1.5-mile-long loop with interpretive signs just north of exit 2 for Rabbit Valley. (See map on page 98.) The trail passes rocks containing *Ankylosaur*, *Camarasaurus*, and *Diplodocus* fossils and skirts the Mygatt-Moore dinosaur quarry, where scientists are still actively unearthing bones and fragments of fossils.

When these dinosaurs were living, they congregated along streams and swamps where food was available in the otherwise semiarid region. These quiet settings offered ideal conditions for the preservation of delicate bones and skeletons because occasional floods would rapidly cover dead animals with sediment before they could be scavenged or destroyed.

The fossilized vertebrae of a Diplodocus *dinosaur are exposed in the Jurassic Morrison Formation along the Trail Through Time.* —Marli B. Miller photo

Morrison Formation along the Trail Through Time. —Marli B. Miller photo

Aerial view of Colorado National Monument showing the layers bending down toward the Grand Valley. Rim Rock Drive snakes along the top edge of the high cliffs in the Wingate Sandstone.
—Marli B. Miller photo

41 Colorado National Monument
The Uplifted Uncompahgre Plateau

Colorado National Monument, on the edge of the Uncompahgre Plateau south of Grand Junction, exhibits towering spires, pinnacles, and fins—all products of erosion. The Uncompahgre Plateau displays many of the same rocks that dazzle visitors at more famous national parks such as Canyonlands. The Redlands fault, which runs along the base of the prominent red cliffs south of Grand Junction, lifted the rocks of the monument relative to those of the valley during the Laramide orogeny. The fault didn't create today's landscape, it merely brought more resistant rocks to the surface. The Colorado and Gunnison Rivers, which join in Grand Junction, eroded the Grand Valley in the softer rock north of the fault. As the monument's rocks were uplifted, they bent rather than broke in a few places, draping down over the fault.

The 1.7-billion-year-old Black Canyon Group—a series of gneisses, schists, and granites—are exposed along the fault and at the base of deep canyons in the monument, where its contact with overlying rocks is the Great Unconformity. The Proterozoic bedrock, exposed when the Ancestral Rockies were uplifted in Pennsylvanian time and all the Paleozoic sedimentary rocks were eroded, is unconformably overlain by the much younger 210-million-year-old Chinle Formation. This classic red bed unit from Triassic time is preserved

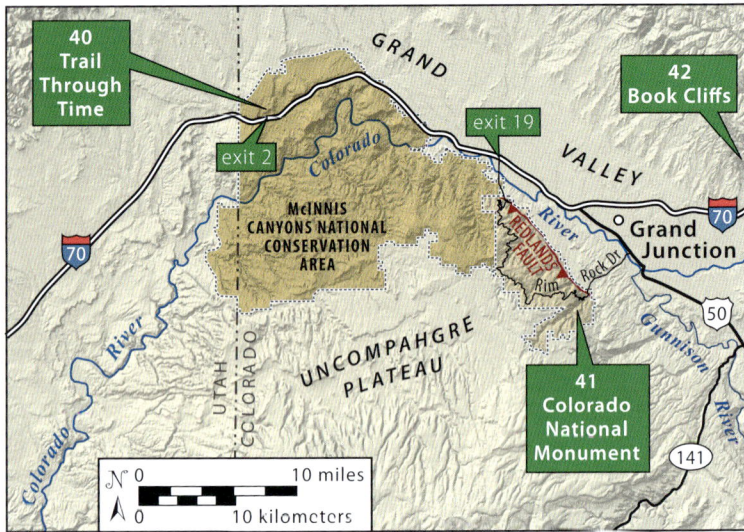

Visitors to Colorado National Monument can loop through the park on Rim Rock Drive.

in great thickness and striking red color across much of the Colorado Plateau.

In the monument, the massive, red vertical cliffs, many rising 350 feet and higher, are the Jurassic-age Wingate Sandstone, a sand dune deposit. Overlying the Wingate are more Jurassic units—the Kayenta, Entrada, Wanakah, and Morrison Formations—that were also mostly deposited on land. Much of the park's 23-mile-long Rim Rock Drive is in the Kayenta Formation, directly above and overlooking the erosional features in the Wingate cliffs. Sandstone cliffs of the Entrada rise above the road. The upper stratigraphic layers are the Cretaceous Burro Canyon Formation, a mainly sandstone unit that features wildly colorful purple and green mudstone layers.

Window Rock is Wingate Sandstone. Note the cross beds in the lighter rock at the upper left.
—National Park Service photo

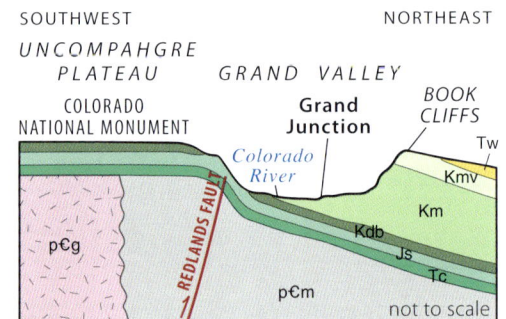

Tw	Wasatch Formation (Paleocene to Eocene)
Kmv	Mesaverde Group (Cretaceous)
Km	Mancos Shale (Cretaceous)
Kdb	Dakota Group and Burro Canyon Formation (Cretaceous)
Js	Wingate, Kayenta, Entrada, Wanakah, and Morrison Formations (Jurassic)
Tc	Chinle red beds (Triassic)
pЄg	granite (Middle Proterozoic)
pЄm	metasedimentary rocks; includes Black Canyon Group (Early Proterozoic)

Movement on the Redlands fault bent the overlying sedimentary rocks, and the Grand Valley was eroded in the soft Mancos Shale.

The Wingate Sandstone of Jurassic age forms Independence Monument and the massive, resistant walls of Monument Canyon in Colorado National Monument. Below it lies the bright-red Chinle Formation. The dark rock at the canyon bottom is the Proterozoic Black Canyon Group. On the far side of Grand Valley in the distance, you can see the Book Cliffs, at the same elevation as the top of the Wingate in the national monument but formed in younger Cretaceous rock.
—Marli B. Miller photo

Layers of the Wingate Sandstone bend sharply above the Redlands fault, which is hidden in the subsurface.
–Marli B. Miller photo

99

Mt. Garfield, at the southern end of the Book Cliffs, is primarily a loose slope of Mancos Shale capped by Mesaverde Group sandstone. —Magdalena S. Donahue photo

42 Book Cliffs
Hydrocarbons in the Piceance Basin

Like the Denver Basin on the east side of the Rockies, the Piceance Basin in western Colorado formed as the Rocky Mountains rose during the Laramide orogeny, from 75 to 45 million years ago. Though not a basin now, this large downwarp between the Rockies to the east and the Uncompahgre uplift to the west collected sediment as the rising mountains eroded. The new uptick in sediment accumulated on top of an already thick sequence deposited in the Western Interior Seaway of Cretaceous time, resulting in 12,000 feet of sediment. As in most thick collections of sediment, the remains of plants and other life transformed into natural gas, oil, and coal as the Piceance Basin became compressed and heated. The sediments were further squeezed by regional tectonics that produced folds in the layered rocks. Folds often trap rising oil and gas beneath impervious layers of rock.

Natural gas, oil shale, and coalbed methane have been harvested from the Paleocene-to-Eocene-age Green River and Wasatch Formations and the Late Cretaceous Mesaverde Group. The region was highly productive in the late twentieth century, then experienced a drop in production. Production recently expanded because the fracking of oil shales increased the economic viability.

The Piceance Basin is rimmed by the resistant Mesaverde Group sandstones, which cap the Grand Hogback on the basin's east side and the Book Cliffs on the basin's south and west sides. You can get an up-close look at the Book Cliffs from the Mt. Garfield trail at the northeast edge of Grand Junction. The Book Cliffs may be so named because their many layers look like the pages in a book. The Mesaverde sandstones protect the underlying Mancos Shale, which erodes into badlands at the cliff's base.

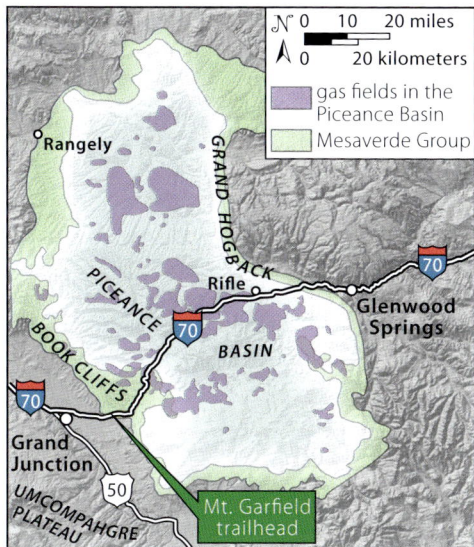

The Piceance Basin collected sediment during the Laramide orogeny and later trapped natural gas.

SOUTHWEST   NORTHEAST

Tgr	Green River Formation (Eocene)
Tw	Wasatch Formation (Paleocene to Eocene)
Kmv	Mesaverde Group (Cretaceous)
Km	Mancos Shale (Cretaceous)
	older rocks

The Mesaverde Group sandstones of the Book Cliffs protect the underlying Mancos Shale from erosion.

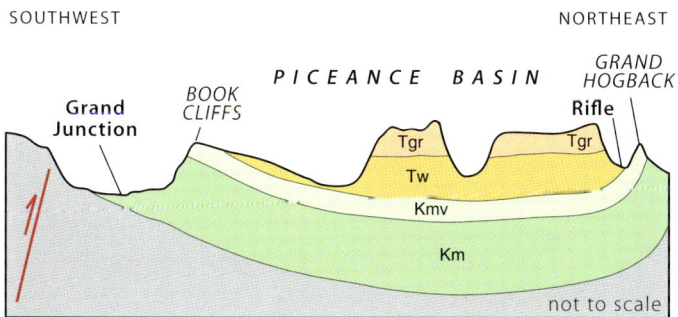

The Book Cliffs extend for miles across western Colorado north of Grand Junction. —Marli B. Miller photo

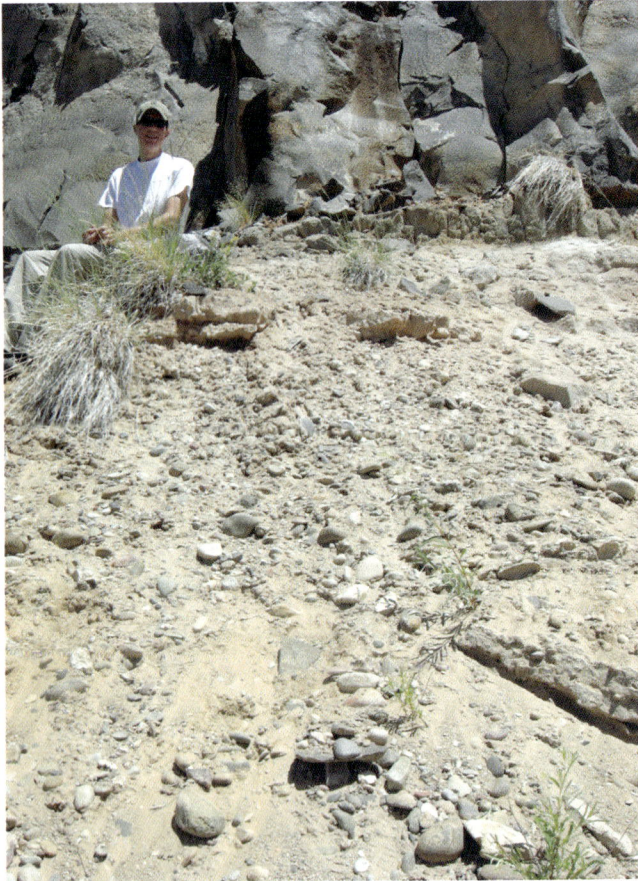

Close-up shows massive dark-gray basalt overlying river gravels with rounded cobbles. —Andres Aslan photo

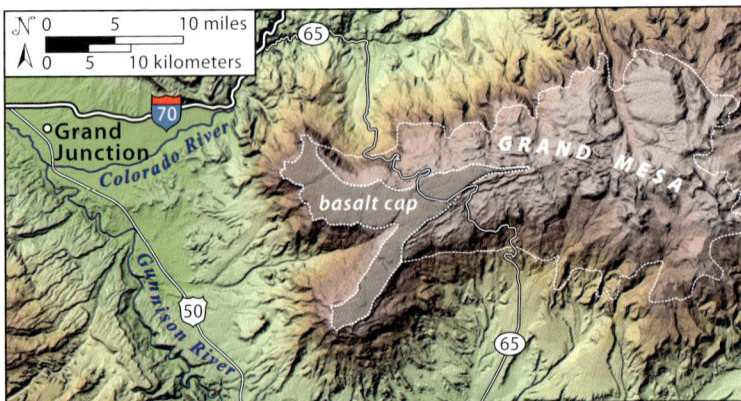

The flat cap of Grand Mesa is basalt that flowed down a river valley 10 million years ago.

43 Grand Mesa
Basalt That Filled a Valley

Grand Mesa, the largest flat-topped mountain in the world, rises to 11,000 feet in elevation above the Colorado River near Grand Junction. Its cool climate, forested meadows, and alpine lakes are a world away from the dry adobe desert at its base. If you want to escape the summer heat, drive across the mesa on CO 65.

The mesa is an excellent example of inverted topography, which means that what is at its top was once the low spot in the topography. At the top of the mesa is basalt that erupted from vents in the Grand Mesa volcanic field to the east around 10 million years ago. Like all liquids, fluid basalt will find the lowest points in the landscape, pooling in and flowing down riverbeds. More than twenty lava flows stacked up, filling in irregularities and cooling into a very flat surface.

Because the ancient riverbed was filled and dammed with volcanic materials, the river changed course and continued to erode the surrounding landscape, removing vast quantities of Mesozoic and Cenozoic sedimentary rock. The more resistant basalt remained intact, resisting erosion and shielding the softer sedimentary layers beneath it. Now, after 10 million years of uplift and erosion, Grand Mesa stands high in the landscape. Gravels and cobbles deposited by the former river are preserved beneath the lowermost basalt flows, now visible at about 10,000 feet elevation! Features in these river deposits tell us that the river flowed west toward the Colorado Plateau.

Erosion continues to eat away at the sides of the mesa as softer rock beneath the lava cap is eroded. The steep slopes of Grand Mesa are in constant battle with gravity, and landslides are common.

Basalt at the top of Grand Mesa directly overlies ancient Colorado River gravels that in turn overlie light-colored, soft sedimentary rocks of the Wasatch Formation and Mesaverde Group. —Andres Aslan photo

Miocene basalt caps Grand Mesa. Beneath the basalt lies a sequence of Eocene Green River Formation (vegetation covered), Cretaceous Mesaverde Group (orangish cliff bands in middle third), and Mancos Shale (bottom). —Marli B. Miller photo

44 Unaweep Canyon
The Mystery of a Missing River

Unaweep Canyon, a half-mile-deep, steep-walled gorge that cuts across the Uncompahgre Plateau, has been a geologic mystery for centuries. This enormous, 4-mile-wide canyon contains two tiny streams, one flowing to the east and one flowing to the west. The small streams we see today are not nearly powerful enough to have carved such a large canyon, and the canyon drains in two directions!

Some early workers considered Unaweep Canyon to be carved by glaciers because of its cross-sectional U shape. This profile, however, is deceptive because the canyon is deeply filled with river-transported material that hides the true river-eroded V shape of the canyon bottom.

Current research suggests that Unaweep Canyon was once a through-flowing canyon that housed a large tributary to the Colorado River, perhaps the ancestral Gunnison or even the Colorado River itself. This river may have excavated a preexisting canyon that was originally carved in Paleozoic time, then filled in with sediment. Regardless of its controversial origin, Unaweep Canyon was abandoned when its river found an easier path. Another Colorado River tributary working its way south toward the Gunnison drainage basin through easily erodible Mancos Shale may have captured the Unaweep stream, a process that geologists call stream piracy.

Unaweep Canyon looks U-shaped, but we are actually seeing just the upper part of a V-shaped canyon. Quaternary gravels fill the lower part. The red sedimentary walls are Mesozoic in age and overlie Proterozoic basement rock. —Magdalena S. Donahue photo

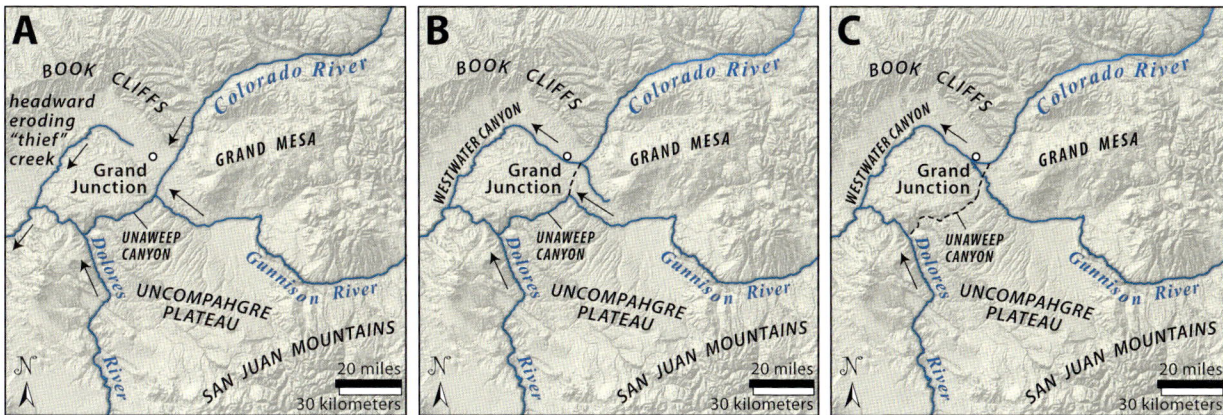

The Colorado River probably flowed through Unaweep Canyon (A) until a headward-eroding stream in Westwater Canyon captured the Colorado (B) upstream from its junction with the Gunnison River. As the Grand Valley was further eroded in the Mancos Shale, the Gunnison River (C) eventually joined the Colorado River, and Unaweep Canyon was abandoned. —Modified from Aslan and others, 2014

The Wingate Sandstone (cliffs of reddish rock) and the Chinle Formation (hidden by trees) overlie the Proterozoic basement rock of Unaweep Canyon. Note the pegmatite veins in the basement rock. Inset shows Proterozoic metasedimentary rock with a white pegmatite vein at left. —Marli B. Miller photos

When viewed from the air, the Black Canyon forms an impressive gash across an otherwise flat landscape. —Marli B. Miller photo

45 Black Canyon of the Gunnison
Youthful Canyon Exposes Ancient Rock

The Black Canyon of the Gunnison combines extreme depth, narrowness, and sheer walls to be one of the most spectacular canyons in North America. It is so deep and narrow that shadows make the canyon walls look dark or black, even on a sunny day. The canyon is narrowest at the Narrows, measuring 40 feet side to side at river level from which the canyon walls rise nearly vertically, with the narrowest rim width of 1,100 feet at Chasm View lookout. The greatest depth is 2,722 feet at Warner Point. You would never guess that this deep canyon exists as you approach the south rim through gentle hills of eroding, dull-black Mancos Shale.

The Black Canyon is carved into basement rocks, largely metamorphic gneiss and schist intruded by younger granitic rock. The 1.4-billion-year-old granitic rock, a quartz monzonite, also forms much of the canyon walls visible from park overlooks. The gneiss of the canyon is beautifully layered, reflecting its original sedimentary origin, which has since undergone significant metamorphism. The layering is also no longer horizontal, having been contorted during metamorphism and faulted during uplift. From one of the most famous lookouts in the national park, you can see the Painted Wall, the tallest exposed cliff face in Colorado. The iconic rock "dragons" that stream across it are large pink pegmatite dikes that formed when magma was injected into cracks and fractures in the existing gneiss and schist. The newly added magma cooled slowly, forming enormous crystals of pink feldspar and white muscovite mica, some of which are up to 6 feet long.

The entire region was uplifted during the Laramide orogeny. Initial carving by the early Gunnison River began in the Oligocene Epoch as mountain runoff was funneled into a single drainage by volcanoes erupting in the West Elk Mountains

106

to the north and the growing San Juan volcanic field to the south. The river became entrenched as it carved through easily eroded volcanic debris and underlying Mesozoic sedimentary rock layers. As the Sawatch Range continued to rise to the east in the river's headwaters, the Gunnison continued to cut downward. Much of this narrow canyon has been carved in the last approximately 1.5 million years during the enormous flows of meltwater of the Pleistocene ice age.

The modern Gunnison River is dammed upstream of the canyon, with seasonal water releases only mimicking the once powerful river. At the steepest part of the Gunnison River, where it flows past the Painted Wall and Chasm View, the river drops at a rate of 53 feet per mile. The massive hard granite here forms a knick zone, a place where a river's gradient increases.

Accessing the Gunnison River within the main canyon is difficult and requires a permit from the visitor center. Most "trails" are steep, poison oak–infested gullies down which an advanced hiker can scramble over 1,000 feet from the upper rims.

The black Proterozoic gneiss and schist of the Painted Wall is crosscut by lighter-colored pegmatites that were injected into the bedrock later in the Proterozoic Eon. —Magdalena S. Donahue photo

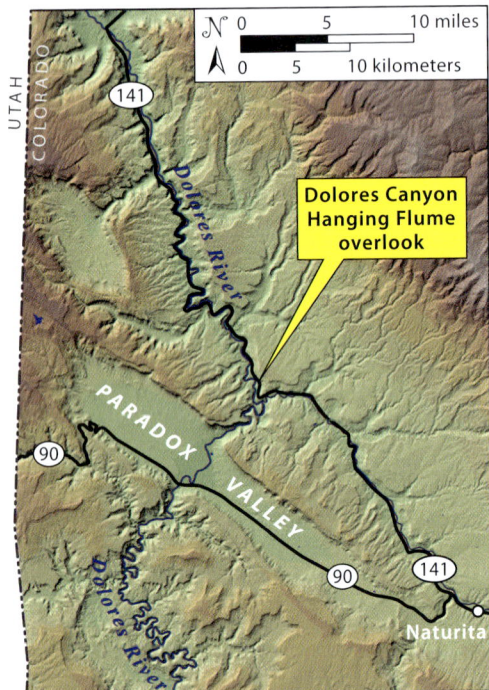

The Dolores River began carving its canyon before the Paradox Valley emerged.

46 Dolores Canyon
Hanging Flume Steals Show from a Geologic Paradox

The Hanging Flume, a wooden chute used to transport water, was built between 1887 and 1891 by the Montrose Placer Mining Company. Along most rivers, water can be diverted into ditches, but the sheer cliffs that rise straight up from the Dolores and San Miguel Rivers required that a 12-mile-long flume be built about 75 feet above the river to carry water to the Lone Tree Placer Mine for hydraulic mining. The three-sided flume is 3 feet wide and was able to support a trough 4 feet deep, capable of transporting 80 million gallons of water per day. Crews who built the flume were suspended from the canyon rim. Iron rods were drilled into the canyon's sandstone walls, upon which a horizontal crosspiece was fitted, supporting the flume channel above it. Each of these pieces—metal and lumber—was uniquely shaped and fitted to the specific geometry of the canyon wall at that spot.

This massive engineering feat required enormous financial resources. Wagons transported hundreds of thousands of board feet of lumber to this remote region, and on-site forges were built to individually shape each iron rod. The flume operated for only three years because it was not profitable. The Hanging Flume is visible from an overlook 21.5 miles north of Naturita on CO 141, the Unaweep-Tabeguache Scenic and Historic Byway. The flume is listed on both the State and National Register of Historic Places.

The Dolores River cuts directly across the Paradox Valley rather than following it. Had the valley existed before the river cut its canyon, the water would surely have just flowed down the valley. But the Paradox Valley did not exist then. The valley, which is aligned along an anticline, or domed-up fold, formed relatively recently and quickly as weathering agents exploited cracks and then reached an easily erodible mass of salt.

Close-up view of the Hanging Flume. —Marli B. Miller photo

108

The steel remnants of the Hanging Flume (arrow) perch above the Dolores River, drilled into the sheer Wingate Sandstone. Above the Wingate lie ledges of the Kayenta Formation, and above that you can see more cliffs of the Entrada Sandstone. As the river incised into bedrock, it became trapped in its meandering path, not able to change course, only to carve a deeper canyon. View from the overlook on CO 141. —Marli B. Miller photo

View looking east across the Paradox Valley toward the Dolores River canyon, the break in the cliffs on the far side of the valley. CO 90 at right. —Steven Baltakatei Sandoval photo, Creative Commons license SA 4.0

47 Four Corners
Natural Gas in the San Juan Basin

The Four Corners is the only location in the United States where the borders of four states intersect: Colorado, New Mexico, Arizona, and Utah. The arid location offers sweeping sedimentary vistas and jagged protruding volcanics, but we will look at what is not visible—valuable oil and natural gas beneath the surface. The petroleum products are in the San Juan Basin, a deep depression that began filling with sedimentary rock about 160 million years ago. Coal, oil, and natural gas formed over time as the sediments became deeply buried and heated. The majority of the basin lies in northwest New Mexico, but it extends north into southwest Colorado, where the state's largest coalbed methane field is located.

The basin is an asymmetrical syncline with more steeply inclined rock layers on the northeast side and a more gradual dip on the southwest side. The downwarping of the crust resulted from the collision and subduction of the Farallon Plate with the western margin of North America, beginning during the late Cretaceous Period. Production wells, which penetrated more than 14,000 feet deep before reaching Proterozoic basement rocks, reveal the thick sequence of

Late Paleozoic and early Mesozoic rocks just north of Durango are inclined southward into the San Juan Basin. —Marli B. Miller photo

SAN JUAN BASIN

	Kpc	Pictured Cliffs Sandstone
	Kl	Lewis Shale
CRETACEOUS	Kmv	Mesaverde Group
	Kmvp	Point Lookout Sandstone
	Kd	Dakota Group

Oil and natural gas are produced from sedimentary rocks in the San Juan Basin.

sedimentary rock in the basin. The oldest sedimentary rocks present above the basement rock were deposited in the sea, beginning in Mississippian time with the Leadville Limestone. Later, in Permian time, the bright-red Cutler Group was deposited on land as the Uncompahgria uplift of the Ancestral Rockies was eroding. During this time, layer after layer of sand and mud began to accumulate in the deepening basin. Mesozoic rock units within the basin include the Triassic Chinle Formation and Jurassic Morrison Formation. In Cretaceous time, the Western Interior Seaway flooded the region with shallow seas, depositing the Lewis Shale and Pictured Cliffs Sandstone. The seaway retreated, leaving a swampy, flooded landscape that eventually produced the coal-rich Fruitland Formation and Kirtland Shale. While the San Juan Basin was warping down, the Rocky Mountains were growing to the east during the Laramide orogeny. Fine beds of ash within the Paleocene-age Ojo Alamo, Animas, and Nacimiento Formations record volcanic activity of the Laramide orogeny.

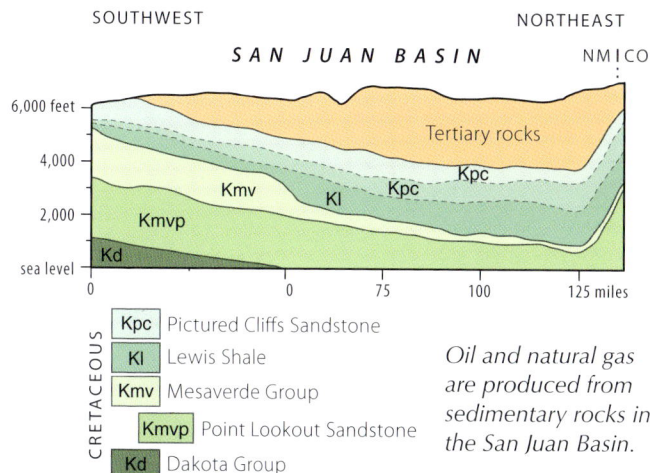

The northern edge of the San Juan Basin extends north into Colorado and the Four Corners region.

Oil derricks dot the surface of the San Juan Basin in Colorado and New Mexico. —Kevin Hobbs photo

111

The vertical cliffs of Point Lookout Sandstone, bathed in golden sunset light, stand vertically above slopes of Mancos Shale. Mancos Shale also floors the valley bottom to the west. —Magdalena S. Donahue photo

48 Mesa Verde National Park
Ancient Ruins in Cliff House Sandstone

Mesa Verde lies in a transition zone between the Colorado Plateau to the west and the Rocky Mountain region to the east. The once flat-lying sedimentary rocks of the park are tilted slightly southward during the Laramide orogeny as part of the northern edge of the San Juan Basin to the east. The area has been eroded into *cuestas*, the technically correct name for tilted mesas.

The sedimentary units visible in Mesa Verde National Park were deposited during the Cretaceous Period, during a temporary retreat of the Western Interior Seaway. The Mancos

Shale, deposited within the sea, is capped by dramatic sandstone cliffs of the Mesaverde Group, deposited along the edge of the sea. The three formations of the Mesaverde Group, from bottom (older) to top (younger), are the Point Lookout Sandstone, the Menefee Formation, and the Cliff House Sandstone. The Point Lookout Sandstone, former beach deposits from a shallowing sea, sits directly atop the Mancos, causing a dramatic shift in topography from easily erodible slopes of shale to impressive cliffs of massive, buff-colored sandstone. The more resistant sandstone serves as a robust caprock, sheltering the shale below it. The Menefee Formation, deposited in swamps, displays coal seams and shaley layers, as well as leaf impressions and tree fossils.

The capping Cliff House Sandstone is so named because of the proclivity for ancestral peoples to build their large houses in this unit. They inhabited the cliff houses until AD 1250. Habitations were often built within alcoves eroded into the relatively soft layers of the Cliff House Sandstone. These softer sections host many natural springs because groundwater tends to flow along and emerge from this interface. The presence of water prompts the weathering and accelerates erosion of rocks along horizontal spring seeps, forming natural alcoves. Inhabitants were both sheltered by and provided with water from the same location in the stratigraphy!

The Mancos Shale, the gray-black mudstone that forms the lower hillslopes and valley floor, contains volcanic ash, which alters to bentonite, an expanding, sticky clay. The Mancos Shale type section—the area in which the unit was first officially described by geologists—is near the town of Mancos along US 160 about 6 miles east of Mesa Verde National Park. The name *Mancos* is Spanish for "one-handed" or "lame," likely a reference to the incredible miring effect the unit has when wet. In this region the Mancos Shale is 2,000 to 4,000 feet thick, and certain layers contain fossil ammonites, mollusks, shark teeth, skates, and rays. The dark color of the shale results from its deposition at the bottom of the sea in an oxygen-poor environment.

The cliff dwellings of Mesa Verde, world-renowned archeological sites, are in alcoves in the Cliff House Sandstone. The black streaks above the cliffs show the path where water has seeped down the cliff face, allowing for bacteria to grow and interact with manganese minerals. —Marli B. Miller photo

49 Bakers Bridge North of Durango
Great Unconformity along the Animas River

You can ride the Durango & Silverton narrow-gauge railroad through the Animas River Canyon, but it doesn't stop at or even come close to Bakers Bridge, where you can see a great example of the Great Unconformity along the Animas River just north of Durango. The 1.7-billion-year-old Bakers Bridge Granite is overlain by marine sandstones of the Ignacio Quartzite of Paleozoic age. The Ignacio Quartzite, once beach sands along a rocky granitic coast, records an early transgression, or

sea level rise, across southern Colorado. The exact age of the Ignacio Quartzite is imprecise; a single brachiopod fossil resembles those found in mid to late Cambrian rocks, but recent finds of fish scale fossils have led some scientists to think the unit is Late Devonian age, or 460 million years old. Either way, more than 1 billion years passed between the cooling of the granite deep within Earth's crust and the deposition of the sand when that same granite was at Earth's surface.

To reach Bakers Bridge, drive 14.3 miles north of US 160 in Durango on US 550 and, just north of Pinkerton Hot Springs, turn right (toward the river) on County Road 250. Look left at the first bridge you come to, over a small slough that fills a former channel of the river. The unconformity is visible on the hillside on the northwest side of this bridge. The reddish-brown sandstones of the Ignacio Quartzite are strongly cemented by silica

View looking upstream from Bakers Bridge. Here, the Animas River flows through a channel in the Bakers Bridge Granite. Note the vertical joints, or fractures, in the granite. —Marli B. Miller photo

US 550 and the Durango & Silverton Railroad diverge near Bakers Bridge north of Durango.

The unconformity (dashed black line) is the contact between the older Bakers Bridge Granite and the younger Ignacio Quartzite. Notice how the contact is irregular. —Marli B. Miller photo

Close-up of Ignacio Quartzite showing some small-scale cross beds. —Marli B. Miller photo

and form beautiful, layered cliffs. If you look closely at the unconformity itself, you can see that it's uneven, a remnant of the ancient landscape on which the sand was deposited. Furthermore, the granite on the southeast side of the bridge lies at a noticeably higher elevation than the unconformity, to indicate even greater variation in the ancient landscape.

The second bridge crosses the Animas River where it eases through a narrow slot in the granite. Bakers Bridge is named for Charles Baker, a prospector who built the first log bridge in this location in 1860. Returning to the Animas valley after the Civil War ended, Baker was killed by Utes. Legend has it that his fortune in gold is buried somewhere near the vicinity of the bridge named for him. The site was also the location for scenes of Robert Redford and Paul Newman jumping from cliff to river in the movie *Butch Cassidy and the Sundance Kid*.

The upper kiva site and some of its masonry. The Chimney Rock spires are in the background. —Marli B. Miller photo

50 Chimney Rock National Monument
Spires in Pictured Cliffs Sandstone

Chimney Rock, about 25 miles west of Pagosa Springs, has long been recognized as an important archaeological site with ancient homes and ceremonial buildings of the Ancestral Puebloans. Chimney Rock and nearby Companion Rock are formed of the 70-million-year-old Pictured Cliffs Sandstone. Approximately 300 feet thick, the unit is formed of marine sandstones with small layers of interbedded shales deposited during the final retreat of the Western Interior Seaway. The Pictured Cliffs Sandstone overlies the Lewis Shale, deposited in the sea when it was deeper.

The pinnacles provide an unusually striking example of differential erosion in which the overlying sandstone erodes much more slowly than the underlying shale. Weak rocks erode more easily and more quickly, forming canyons and basins, while more resistant rocks remain intact, forming hills, ridges, even mountain ranges. In this location, the Lewis Shale and Fruitland Formation (sandstone, shale, and coal) have eroded more quickly, leaving the resistant Pictured Cliffs Formation standing high in the landscape. Eventually, the pinnacles of Chimney Rock will become unstable and topple as the surrounding base rocks erode. The ancestral Piedra River and its tributaries carved deep valleys as they carried huge amounts of meltwater from receding glaciers at the end of the last ice age. Erosion continues today but not nearly as fast as in the past.

The archaeological sites lie on the narrow ridgeline to the southwest of the spires. Some scholars think the site was built in the high location for astronomical reasons, and indeed, during certain years and lunar cycles, the moon rises between the two spires. The site was abandoned sometime between AD 1125 and AD 1130.

The tall spires of Chimney Rock consist of Pictured Cliffs Sandstone, which is far more resistant to erosion than the underlying Lewis Shale.
—Marli B. Miller photo

GLOSSARY

accreted. Something that has been added onto, such as an accreted terrane, which has been added to the North American continent through plate motions.

alluvial fan. A gently sloping, fan-shaped accumulation of sediments deposited by a stream where it flows out of a narrow valley onto a wider, flatter area.

andesite. A medium- to dark-colored volcanic rock that is between basalt and rhyolite in silica content.

anticline. A fold with the oldest rock in the core; most anticlines have limbs that dip away from the core.

aquifer. Any geologic unit (either sediment or rock) with interconnected pores that store groundwater.

ash. Tiny particles of volcanic glass blown into the air during volcanic eruptions.

ash-flow tuff. The rock formed from the consolidation and compaction of an ash-flow deposit.

asthenosphere. The zone of somewhat malleable rock beneath the lithosphere. It is the zone over which the lithospheric plates move.

badlands. A type of topography characterized by a great number of intricate streams separated by narrow ridges with steep slopes. Badlands develop in areas with little vegetation or poorly consolidated rocks.

basalt. A dark-colored volcanic rock that contains less than 52 percent silica.

basement. The deepest crustal rocks of a given area. They are typically igneous or metamorphic rock.

batholith. A mass of coarse, granular granitic igneous rock exposed over an area greater than about 30 square miles and consisting of two or more plutons.

bedding. The layering as seen in a sedimentary rock. A single layer is called a **bed**. When different rock types are interlayered with each other, they are described as **interbedded**.

bedrock. Rock that remains in its place of origin and has not been moved by erosional processes.

biotite. A black to dark-brown, shiny, iron-rich mica mineral that breaks into thin, flexible sheets. It is a common component of granite, gneiss, and schist.

braided stream. A stream with a floodplain composed of many interconnected, shifting channels separated by coarse sand or gravel bars.

breccia. A rock consisting of angular fragments.

caldera. A steep-walled, subcircular depression in a volcano, at least 1 mile across, that formed by collapsing into an emptied or partially emptied magma chamber below.

caprock. A more resistant rock at the surface that protects the underlying rocks from erosion.

carbonate. A class of sedimentary rocks, such as limestone or dolomite, that are formed by the combination of atoms of calcium or magnesium with carbon and oxygen.

chert. An extremely fine-grained sedimentary rock made of silica.

chlorite. A green, platy mineral characteristic of low-temperature metamorphism of rocks with iron and magnesium.

cirque. A bowl-shaped basin on a mountain, usually where the head of a glacier once existed.

clast. A grain or fragment of a rock. Clastic rock is sedimentary rock composed of broken fragments, such as sand grains, derived from preexisting rocks.

clay. A sedimentary particle with a grain size less than 0.004 millimeter in diameter.

claystone. A fine-grained sedimentary rock composed mostly of clay minerals but lacking the fine layering typical of shale.

coal. An organic-rich, dark-colored to black rock that burns. Coal forms from the compaction and long-term, low-temperature heating of plant material.

coarse grained. Said of sedimentary rocks that have clasts, or particles, that are relatively large, usually averaging 2 millimeters in diameter or larger. Also said of igneous rocks with relatively large crystals.

columnar jointing. The fracturing in a lava flow that causes the flow to break into columns.

conglomerate. A sedimentary rock composed of particles that exceed 2 millimeters in diameter.

continental shelf. The gently inclined part of the continental landmass between the shoreline and the more steeply inclined continental slope.

craton. A part of the Earth's crust, usually in the interior of a continent, that has not been pervasively deformed for at least 1 billion years.

cross bedding. A layering in sedimentary rock that forms at an angle to horizontal.

crust. The uppermost layer of Earth. Continental crust consists mainly of an igneous and/or metamorphic basement overlain by sedimentary and volcanic rock. Oceanic crust consists of basalt and gabbro, an intrusive rock.

dacite. A volcanic rock that is intermediate in silica content between andesite and rhyolite.

delta. A nearly flat accumulation of clay, sand, and gravel deposited in a lake or ocean at the mouth of a river.

dike. An intrusive body that cuts across layering in the host rock.

dolomite. A sedimentary rock composed of the mineral dolomite, a calcium-magnesium carbonate. Most dolomite forms when calcium in limestone is replaced by magnesium.

dune field. An extensive deposit of sand.

earthquake. The shaking caused by abrupt displacement along a fault.

erosion. The movement or transport of weathered material by water, ice, wind, or gravity.

evaporite. A nonclastic sedimentary rock typically formed by the partial or total evaporation of brine.

fault. A fracture or zone of fractures in Earth's crust along which blocks of rock on either side have shifted.

feldspar. The most abundant rock-forming mineral group. Makes up 60 percent of Earth's crust and contains calcium, sodium, or potassium with aluminum silicate.

fine grained. Said of sedimentary rocks that have clasts, or particles, that are relatively small, usually averaging less than 2 millimeters in diameter. Also said of igneous rocks with relatively small, hard-to-see crystals.

fin. A tilted layer of sedimentary rock that projects above the surrounding landscape.

floodplain. The portion of a river valley that is built of sediments deposited when the river overflows its banks during flooding.

formation. A body of sedimentary, igneous, or metamorphic rock that can be recognized over a large area. It is the basic stratigraphic unit in geologic mapping. A formation may be part of a larger group and may be broken into members.

fossils. The remains, imprints, or traces of plants or animals preserved in rock.

glacial. A term pertaining to a glacier.

glacial till. An unsorted mixture of silt, sand, and gravel left by a melting glacier.

glacier. A large and long-lasting mass of ice on land that flows downhill in response to gravity.

gneiss. A coarse-grained metamorphic rock having a banded or striped appearance. Gneisses form at higher metamorphic grades.

granite. A light-colored, coarse-grained igneous rock with a silica content that exceeds 66 percent. It is the intrusive equivalent of rhyolite. The term **granitic** pertains to an igneous rock that resembles and approximates the chemical composition of a granite.

granodiorite. A coarse-grained, light-colored igneous rock that looks very similar to granite but contains more plagioclase feldspar and less potassium feldspar. It is the intrusive equivalent of dacite.

groundwater. The subsurface water contained in fractures and pores of rock and soil.

group. Two or more formations that occur together.

gypsum. Hydrous calcium sulphate, a soluble mineral that forms as water evaporates.

hogback. A long, linear ridge with steep sides and a narrow crest. Hogbacks are so named because they resemble the back of a hog.

hoodoos. Tall erosional pinnacles, most commonly in loose, poorly cemented sediments.

hydrocarbon. A gas or liquid, such as oil and natural gas, consisting of carbon and hydrogen.

hydrothermal. Related to hot water below Earth's surface.

igneous rock. A rock that solidified from the cooling of molten magma or lava.

impermeable. Having a texture that does not permit water to move through. Clay is typically considered a relatively impermeable sediment.

incision. A process whereby a stream deepens its channel, particularly in relation to uplift. A river flowing in a deep, narrow valley or canyon is said to be **incised**.

intrusive rocks. Igneous rocks that cool from magma beneath the surface of Earth. The body of rock is called an **intrusion**.

laccolith. A mushroom-shaped igneous intrusion that intrudes parallel to sedimentary rock layers and bows them upward.

lahar. A volcanic mudflow deposit.

lava. Molten rock erupted on the surface of Earth.

limestone. A sedimentary rock composed of calcium carbonate precipitated in warm water, aided by biological activity.

lithosphere. The outer rigid shell of Earth that is broken into tectonic plates. On average, continental lithosphere is about 100 miles thick and old oceanic lithosphere is about 60 miles thick.

magma. Molten rock within Earth.

mantle. The part of Earth between the core and the outer crust.

marble. Metamorphosed limestone.

marine. Pertaining to the sea.

metamorphic rock. A rock derived from preexisting rock that has changed mineralogically or texturally, or both, in response to changes in temperature and/or pressure, usually deep beneath Earth's surface.

metamorphism. The recrystallization of an existing rock. Metamorphism typically occurs at high temperatures and often high pressures.

moraine. A mound or ridge of an unsorted mixture of silt, sand, and gravel (glacial till) left by a melting glacier.

mountain building. See **orogeny**.

mudstone. A sedimentary rock composed of mud.

normal fault. A fault in which rocks above the fault move down relative to rocks below the fault. Normal faults result in extension, or stretching of the crust.

orogeny. An event in which mountains rise. During these events, rocks are typically folded, faulted, and/or metamorphosed. Intrusive and extrusive igneous activity often accompanies these events.

oxide. A compound in which oxygen combines with a positively charged ion. Iron oxide is a common example.

pebble. A rounded rock particle 0.16 to 2.5 inches in diameter.

pegmatite. A very coarse-grained granitic rock with large, interlocking crystals.

piedmont. The plains, foothills, or slope at the base of mountains.

plagioclase. A feldspar mineral rich in sodium and calcium. One of the most common rock-forming minerals in igneous and metamorphic rocks.

playa. An ephemeral lake, one that may fill partially with water during wetter months and dry out partially or completely during dry months.

pluton. A body of intrusive igneous rock.

pyroclastic. Volcanic material that becomes broken into small pieces during an explosive eruption.

quartz. A mineral composed entirely of silica; one of the most common rock-forming minerals.

quartzite. A metamorphic rock consisting of tightly interlocked quartz crystals. Originates as a quartz-rich sandstone.

quartz monzonite. A coarse-grained, light-colored granitic rock with more quartz than granodiorite and less alkali feldspar than true granite.

radiometric dating. The calculation of age based on the rate of time it takes for radioactive elements to decay. When it is based on carbon, it is called **radiocarbon dating**.

red bed. Sedimentary rocks that are predominantly red because of the presence of iron oxides.

rhyolite. A typically light-colored volcanic rock with more than 66 percent silica. It is the volcanic equivalent of granite.

rift zone. A strip-like area characterized by crustal extension and normal faulting.

sand. Weathered mineral grains, most commonly quartz, between 0.06 and 2 millimeters in diameter.

sandstone. A sedimentary rock made primarily of sand.

scarp. A line of cliffs produced by faulting or landsliding.

schist. A metamorphic rock that is strongly layered due to an abundance of visible platy minerals.

sedimentary rock. A rock formed from the compaction and cementation of sediment.

shale. A thinly layered rock made of sedimentary particles less than 0.004 millimeter in diameter.

shear. The action caused by the side-by-side sliding of two particles past each other.

silica. The compound silicon dioxide. The most common mineral made entirely of silica is quartz.

sill. An igneous intrusion that parallels the planar structure or bedding of the host rock.

silt. Sedimentary particles larger than clay but smaller than sand (between 0.004 and 0.06 millimeter in diameter).

siltstone. A sedimentary rock made primarily of silt.

stock. A pluton smaller than 40 square miles in area. A cluster of stocks may be large enough to form a batholith.

stratification/stratified. Sequentially layered.

stratigraphic. Pertaining to the sequence of sedimentary rock layers found in a given region.

stratovolcano. A steep-sided volcano, typically made of andesite.

subduction zone. A long, narrow zone where an oceanic plate descends beneath another plate at a convergent boundary.

syenite. A coarse-grained plutonic igneous rock consisting dominantly of orthoclase (potassium-sodium feldspar).

syncline. A fold in which layered rocks have been bowed downward, producing a smile-like profile.

tectonics. The study of regional-scale deformation of Earth's crust.

tectonic plates. Large fragments, or plates, that move slowly over the somewhat malleable asthenosphere, with intense geologic activity at plate boundaries.

terrace. An erosional remnant of a former floodplain standing above the present river.

till. An unlayered and unsorted mixture of clay, silt, sand, gravel, and boulders deposited directly by a glacier.

travertine. A sedimentary rock produced by the precipitation of calcite at springs or in caves.

tuff. A volcanic rock made mostly of consolidated pyroclastic material, chiefly ash and pumice, derived from ash falls or pyroclastic flows. A **welded tuff** is distinctly harder because heat of its particles caused them to weld together.

unconformity. A contact between two rock bodies representing a period of erosion or lack of deposition during a significant length of geologic time.

vein. A deposit of minerals that fills a fracture in rock.

vent. The actual place where volcanic materials erupt. Vents are either eruptive localities on large volcanoes or mark much smaller volcanoes.

volcano. An opening in Earth's crust where magma reaches the surface.

volcanic field. An area covered with volcanic rocks erupted from the same group of volcanoes.

volcanism. Processes involved with the eruption of lava and ash at Earth's surface.

weathering. The physical disintegration and chemical decomposition of rock at Earth's surface.

zeolite. Aluminum-silicate minerals, containing sodium, potassium, or calcium, that commonly form where volcanic rocks or ash interact with groundwater or volcanic fluids.

REFERENCES

General Colorado

Abbott, L., and T. Cook. 2012. *Geology Underfoot along Colorado's Front Range*. Missoula, MT: Mountain Press Publishing Company.

Colorado Geological Survey. 2003. *Messages in Stone: Colorado's Colorful Geology*. Colorado Department of Natural Resources.

Taylor, A. M. 1999. *Guide to the Geology of Colorado*. Golden, CO: Cataract Lode Mining Company.

Williams, F., and H. Chronic. 2014. *Roadside Geology of Colorado*, Third Edition. Missoula, MT: Mountain Press Publishing Company.

Online Resources

Colorado Geological Survey
https://coloradogeologicalsurvey.org

Denver Museum of Nature and Science
www.dmns.org

Rocky Mountain Association of Geologists
They publish *The Mountain Geologist*, a nontechnical quarterly journal for the interested reader.
https://www.rmag.org/

Front Range and Great Plains

2. DEVILS BACKBONE

Braddock, W. A., R. H. Calvert, S. J. Gawarecki, and P. Nutalaya. 1972. *Geologic Map of the Masonville Quadrangle, Larimer County, Colorado*. USGS Geologic Quadrangle Map GQ-832, 1:24,000.

5. FLATIRONS

Ellis, C. H. 1966. Paleontologic age of the Fountain Formation south of Denver, Colorado. *The Mountain Geologist* 3 (4): 155–60.

Gable, D. J. 1980. *Boulder Creek Batholith, Front Range, Colorado*. USGS Professional Paper 1101.

6. I-70 ROADCUT

Weimer, R. J., and L. W. LeRoy. 1987. Paleozoic-Mesozoic section: Red Rocks Park, I-70 road cut, and Rooney Road, Morrison area, Jefferson County, Colorado. In *Centennial Field Guide Volume 2*, Rocky Mountain Section of the GSA, ed. S. S. Beus, p. 335–38.

7. DINOSAUR RIDGE

Modreski, P. J. 2001. Geochemical and mineralogical studies of dinosaur bone from the Morrison Formation at Dinosaur Ridge. *The Mountain Geologist* 38 (3): 111–18.

8. ROXBOROUGH STATE PARK

Bishop, E. M., S. Weaver, and M. B. Miller. 2013. Paleozoic to Mesozoic sedimentary formations in the Colorado Front Range, Roxborough State Park: Documenting geologic landscapes and features through geo-photography. In *Classic Concepts and New Directions: Exploring 125 years of GSA Discoveries in the Rocky Mountain Region*, GSA Field Guide 33, eds. L. D. Abbott and G. S. Hancock, p. 279–94.

Scott, G. R. 1963. *Bedrock Geology of the Kassler Quadrangle, Colorado*. USGS Professional Paper 421-B. Map accompanies the paper as Plate 2.

9. CASTLE ROCK

Keller, S. M., and M. L. Morgan. 2016. *New Paleocurrent Measurements, Clast Population Data, and Age Dates in the Late Eocene Castle Rock Conglomerate, East-Central Colorado: Remapping the Fluvial System, and Implications for History of the Colorado Piedmont and Front Range*. Colorado Geological Survey Open-File Report 16-01.

Koch, A. J., D. S. Coleman, and A. M. Sutter. 2018. Provenance of the upper Eocene Castle Rock Conglomerate, south Denver Basin, Colorado, USA. *Rocky Mountain Geology* 53 (1): 29–43.

Lipman, P. W., M. J. Zimmerer, and W. C. McIntosh. 2015. An ignimbrite caldera from the bottom up: Exhumed floor and fill of the resurgent Bonanza caldera, Southern Rocky Mountain volcanic field, Colorado. *Geosphere* 11 (6): 1902–47.

McIntosh, W. C., and C. E. Chapin. 2004. Geochronology of the central Colorado volcanic field. *New Mexico Bureau of Geology and Minerals Resources Bulletin* 160: 205–37.

10. GARDEN OF THE GODS

Keller, J. W., C. S. Siddoway, M. L. Morgan, and others. 2005. *Geologic Map of the Manitou Springs 7.5-Minute Quadrangle, El Paso and Teller Counties, Colorado*. Colorado Geological Survey Open-File Report 03-19.

Morgan, M. L., C. S. Siddoway, P. D. Rowley, and others. 2003. *Geologic Map of the Cascade Quadrangle, El Paso County, Colorado*. Colorado Geological Survey Open-File Report 03-18.

Ross, M. R., W. Hoesch, S. Austin, J. H. Whitmore, and T. Clarey. 2010. Garden of the Gods at Colorado Springs: Paleozoic and Mesozoic sedimentation and tectonics. In *Through the Generations: Geologic and Anthropogenic Field Excursions in the Rocky Mountains from Modern to Ancient*, eds. L. Morgan and S. Quane, p. 77–93.

Siddoway, C., P. Myrow, and E. Fitz-Diaz. 2013. Strata, structures, and enduring enigmas: a 125th Anniversary appraisal of Colorado Springs Geology. In *Classic Concepts and New Directions: Exploring 125 years of GSA Discoveries in the Rocky Mountain Region*, GSA Field Guide 33, eds. L. D. Abbott and G. S. Hancock, p. 331–56.

Sterne, E. J. 2006. Stacked "evolved" triangle zones along the southeastern flank of the Colorado Front Range. *The Mountain Geologist* 43: 65–92.

11. PAINT MINES

Johnson, K. R., M. L. Reynolds, K. W. Werth, and J. R. Thomasson. 2003. Overview of the Late Cretaceous, early Paleocene, and early Eocene megafloras of the Denver Basin, Colorado. *Rocky Mountain Geology* 38 (1): 101–20.

12. CAVE OF THE WINDS

Gerhard, L. C. 1974. Redescription and new nomenclatures of Manitou Formation, Colorado. *American Association of Petroleum Geologists Bulletin* 58 (7): 1397–1406.

Swett, K. 1964. Petrology and paragenesis of the Ordovician Manitou Formation along the Front Range of Colorado. *Journal of Sedimentary Research* 34 (3): 615–24.

13. PIKES PEAK

Chapin, C. E., S. A. Kelley, and S. M. Cather. 2014. The Rocky Mountain Front, southwestern USA. *Geosphere* 10 (5): 1043–60.

Smith, D. R., R. A. Wobus, J. B. Noblett, D. Unruh, and K. Chamberlain. 1999. A review of the Pikes Peak batholith, Front Range, central Colorado: A 'type example' of A-type granitic magmatism. *Rocky Mountain Geology* 34 (2): 93–116.

14. FLORISSANT FOSSIL BEDS

Cather, S. M., C. E. Chapin, and S. A. Kelley. 2012. Diachronous episodes of Cenozoic erosion in southwestern North America and their relationship to surface uplift, paleoclimate, paleodrainage, and paleoaltimetry. *Geosphere* 8 (6): 1177–1206.

Johnson, K. R., E. Evanoff, K. M. Gregory-Wodzicki, and others. 2001. *Fossil Flora and Stratigraphy of the Florissant Formation, Colorado*. Proceedings of the Denver Museum of Nature and Science, series 1, no. 4.

Larsen, D. 2000. Upper Eocene and Oligocene Lacustrine Deposits of the Southwestern United States, with Emphasis on the Creede and Florissant Formations. In *Lake Basins through Space and Time*, AAPG Studies in

Geology 46, eds E. H. Gierlowski-Kordesch and K. R. Kelts, p. 425–38.

Leonard, E. M., M. S. Hubbard, S. A. Kelley, and others. 2002. High Plains to Rio Grande Rift: Late Cenozoic Evolution of Central Colorado. In *Science at the Highest Level*, GSA Field Guide 3, ed. D. Lageson, p. 59–93.

15. ROYAL GORGE

McIntosh, W. C., and C. E. Chapin. 2004. Geochronology of the central Colorado volcanic field. *New Mexico Bureau of Geology and Minerals Resources Bulletin* 160: 205–37.

Taylor, R. B., G. R. Scott, R. A. Wobus, and R. C. Epis. 1975. *Reconnaissance geologic map of the Royal Gorge Quadrangle, Fremont and Custer Counties, Colorado*. USGS IMAP 869.

16. PICKETWIRE CANYONLANDS

Heckert, A. B., E. J. Sload, S. G. Lucas, and B. A. Schumacher. 2012. Triassic fossils found stratigraphically above 'Jurassic' eolianites necessitate the revision of lower Mesozoic stratigraphy in Picket Wire Canyonlands, south-central Colorado. *Rocky Mountain Geology* 47 (1): 37–53.

Schumacher, B., and M. Lockley. 2014. Newly documented trackways at Dinosaur Lake: the Purgatoire Valley dinosaur tracksite. In *Fossil Footprints of Western North America*. eds. M. G. Lockley and S. G. Lucas, p. 261–69.

17. TRINIDAD LAKE STATE PARK

Barclay, R. S., K. R. Johnson, W. J. Betterton, and D. L. Dilcher. 2003. Stratigraphy and megaflora of a K-T boundary section in the eastern Denver Basin, Colorado. *Rocky Mountain Geology* 38 (1): 45–71.

Izett, G. A. 1987. *The Cretaceous-Tertiary boundary interval, Raton Basin, Colorado and New Mexico and its content of shock-metamorphosed minerals*. USGS Open-File Report 87-606.

18. RADIAL DIKES OF SPANISH PEAKS

Knopf, A. 1936. Igneous geology of the Spanish Peaks region, Colorado. *GSA Bulletin* 47 (11): 1727–84.

Central Rockies

19. CLIMAX MINE

Lipman, P. W., F. S. Fisher, H. H. Mehnert, and others. 1976. Multiple ages of Mid-Tertiary mineralization and alteration in the western San Juan Mountains, Colorado. *Economic Geology* 71 (3): 571–88

Wallace, S. R., N. F. Muncaster, D. C. Johnscon, and others. 1968. Multiple intrusion and mineralization at Climax, Colorado. In *Ore deposits of the United States, 1933–1967*, ed. J. D. Ridge, American Institute of Mining, Metallurgical and Petroleum Engineers, p. 605–40.

20. LEADVILLE MINING DISTRICT

Chapin, C. E. 2012. Origin of the Colorado Mineral Belt. *Geosphere* 8 (1): 28–43.

Emmons, S. F. 1879. *Geology and Mining Industry of Leadville, Colorado*. Reprinted in 2017 by Forgotten Books.

Garcia, R. V. 2011. *Cenozoic intrusive and exhumation history of the West Elk Mountains, southwestern Colorado*. MS Thesis, New Mexico Institute of Mining and Technology.

Tweto, 0., R. H. Moench, and J. C. Reed, Jr. 1978. *Geologic Map of the Leadville 1° x 2° Quadrangle, Northwestern Colorado*. USGS Miscellaneous Investigations Map 1-999, scale 1:250,000.

Wallace, A. R. 1993. *Geologic Setting of the Leadville Mining District, Lake County, Colorado*. USGS Open-File Report 93-343.

21. CHALK CLIFFS OF MT. PRINCETON

Lipman, P. W., and O. Bachmann. 2015. Ignimbrites to batholiths: Integrating perspectives from geological, geophysical, and geochronological data. *Geosphere* 11 (3): 705–43.

Nakai, J. S., A. F. Sheehan, and S. L. Bilek. 2017. Seismicity of the Rocky Mountains and Rio Grande Rift from the EarthScope Transportable Array and CREST temporary seismic networks, 2008–2010. *Journal of Geophysical Research Solid Earth* 122 (3): 2173–92.

Ricketts, J. W., S. A. Kelley, K. E. Karlstrom, and others. 2016. Synchronous opening of the Rio Grande rift along its entire length at 25–10 Ma supported by apatite (U-Th)/He and fission-track thermochronology, and evaluation of possible driving mechanisms. *GSA Bulletin* 128 (3-4): 397–424.

22. YULE MARBLE

McGee, E. S. 1999. *Colorado Yule Marble: Building Stone of the Lincoln Memorial.* USGS Bulletin 2162.

23. MAROON BELLS

Bryant, B. 1969. *Geologic Map of the Maroon Bells Quadrangle, Pitkin and Gunnison Counties, Colorado.* USGS Geologic Quadrangle 788.

Campbell, D. L. 1985. *Gravity and Aeromagnetic Maps of the Maroon Bells–Snowmass Wilderness and Additions, Pitkin and Gunnison Counties, Colorado.* USGS Miscellaneous Field Studies Map 1647-B.

Pecha, M. E., G. E. Gehrels, K. E. Karlstrom, and others. 2018. Provenance of Cretaceous through Eocene strata of the Four Corners region: Insights from detrital zircons in the San Juan Basin, New Mexico and Colorado. *Geosphere* 14 (2): 785–811.

24. THE GROTTOS

Fridrich, C. J., E. DeWitt, B. Bryant, S. Richard, and R. P. Smith. 1998. *Geologic Map of the Collegiate Peaks Wilderness Area and the Grizzley Peak Caldera, Sawatch Range, Central Colorado.* USGS Miscellaneous Investigations Series, Map I-2565.

25. GLENWOOD CANYON

Aslan, A., K. E. Karlstrom, L. J. Crossey, and others. 2010. Late Cenozoic evolution of the Colorado Rockies: Evidence for Neogene uplift and drainage integration. In *Through the Generations: Geologic and Anthropogenic Field Excursions in the Rocky Mountains from Modern to Ancient*, GSA Field Guide 18, eds. L. A. Morgan and S. L. Quane, p. 21–54.

Decker, K. 2010. Evaluation of Hanging Lake, Garfield County, Colorado for its Merit in Meeting National Significance Criteria as a National Natural Landmark. In *Representing Lakes, Ponds and Wetlands in the Southern Rocky Mountain Province*, Colorado Natural Heritage Program, Colorado State University, Fort Collins.

Lazear, G., K. E. Karlstrom, A. Aslan, and S. Kelly. 2013. Denudation and flexural isostatic response of the Colorado Plateau and southern Rocky Mountains region since 10 Ma. *Geosphere* 9 (4): 792–814.

26. GLENWOOD SPRINGS

Karlstrom, K., L. J. Crossey, D. R. Hilton, and P. H. Barry. 2013. Mantle He-3 and CO_2 degassing in carbonic and geothermal springs of Colorado and implications for neotectonics of the Rocky Mountains. *Geology* 41 (4): 495–98.

Kirkham, R. M., J. M. Zook, and M. A. Sares. 1999. *Reconnaissance Field Investigation of Surface-Water Specific Conductance in the Snowmass-Glenwood Springs Area, West-Central Colorado.* Colorado Geological Survey.

27. RIFLE FALLS STATE PARK

Scott, R. B., R. R. Shroba, and A. E. Egger. date. *Geologic Map of the Rifle Falls Quadrangle, Garfield County, Colorado.* USGS Miscellaneous Field Studies Map MF-2341.

28. DILLON PINNACLES

Bove, D. J., K. Hon, K. E. Budding, and others. 1994. *Geochronology and Geology of Late Oligocene through*

Miocene Volcanism and Mineralization in the Western San Juan Mountains, Colorado. USGS Professional Paper 1642.

Lipman, P. W., and W. C. McIntosh. 2008. Eruptive and noneruptive calderas, northeastern San Juan Mountains, Colorado: Where did the ignimbrites come from? *GSA Bulletin* 120 (7-8): 771–95.

29. CRESTED BUTTE

Gaskill, D. L. 1986. *Geologic Map of the Crested Butte Quadrangle, Gunnison County, Colorado.* USGS Geologic Quadrangle 1580. Scale 1:24,000.

Gaskill, D. L., F. E. Mutschler, J. H. Kramer, J. A. Thomas, and S. G. Zahony. 1991.

Geologic Map of the Gothic quadrangle, Gunnison County, Colorado. US Geological Survey Geologic Quadrangle Map GQ-1689.

Robinson, C. H., and P. A. Dea. 1981. Quaternary glacial and slope-failure deposits of the Crested Butte area, Gunnison County, Colorado. In *Western Slope (Western Colorado),* New Mexico Geological Society Fall Field Conference Guidebook, eds. R. C. Epis and J. F. Callender, p. 155–63.

Dowsett, F. R., M. W. Ganster, R. E. Ranta, D. J. Baker, and H. J. Stein. 1981. Geology of the Mt. Emmons molybdenum deposit, Crested Butte, Colorado. In *Western Slope (Western Colorado),* New Mexico Geological Society Fall Field Conference Guidebook, eds. R. C. Epis and J. F. Callender, p. 325–32.

San Juan Mountains and Rio Grande Rift

30. SAN LUIS VALLEY

Hudson, M. R., and V. J. S. Grauch. 2013. Introduction. In *New Perspectives on Rio Grande Rift Basins: From Tectonics to Groundwater,* GSA Special Paper 494, eds. M. R. Hudson and V. J. S. Grauch, p. v–xii.

Lindsey, D. 2010. *The Geologic Story of Colorado's Sangre de Cristo Range.* USGS Circular 1349.

Lipman, P. W. 2007. Incremental assembly and prolonged consolidation of Cordilleran magma chambers: Evidence from the Southern Rocky Mountain volcanic field. *Geosphere* 3: 42–70.

Machette, M. N., D. W. Marchetti, and R. A. Thompson. 2007. Ancient Lake Alamosa and the Pliocene to Middle Pleistocene Evolution of the Rio Grande. In *Rocky Mountain Section Friends of the Pleistocene Field Trip: Quaternary Geology of the San Luis Basin in Colorado and New Mexico,* USGS Open-File Report 2007–1193, eds. M. N. Machette, M-M. Coates, and M. L. Johnson, p. 157–67.

31. GREAT SAND DUNES

Burroughs, R. L. 1971. Geology of the San Luis Hills, south-central Colorado. In *San Luis Basin (Colorado),* New Mexico Geological Society 22nd Annual Fall Field Conference Guidebook, ed. H. L. James, p. 277–87.

Hudson, M. R., and V. J. S. Grauch. 2013. Introduction. In *New Perspectives on Rio Grande Rift Basins: From Tectonics to Groundwater,* GSA Special Paper 494, eds. M. R. Hudson and V. J. S. Grauch, p. v–xii.

Madole, R. F., J. H. Romig, J. N. Aleinikoff, D. P. VanSistine, and E. Y. Yacob. 2008. On the origin and age of the Great Sand Dunes, Colorado. *Geomorphology* 99: 99–119.

32. PAGOSA SPRINGS

Barnes, H. 1953. *Geology of the Ignacio area, Ignacio and Pagosa Springs quadrangles, La Plata and Archuleta Counties, Colorado.* USGS Oil and Gas Investigations Map 138, 1:63,360.

33. WHEELER GEOLOGIC AREA

Lipman, P. W. 2007. Incremental assembly and prolonged consolidation of Cordilleran magma chambers: Evidence from the Southern Rocky Mountain volcanic field. *Geosphere* 3: 42–70.

Steven, T., and P. Lipman. 1976. *Calderas of the San Juan volcanic field, southwestern Colorado*. USGS Professional Paper 958.

34. SLUMGULLION EARTHFLOW

Crandell, D. R., and D. J. Varnes. 1960. Slumgullion earthflow and earthslide near Lake City, Colorado. *GSA Bulletin* 71 (12): 1846.

Crandell, D. R., and D. J. Varnes. 1961. Movement of the Slumgullion earthflow near Lake City, Colorado. In *Short papers in the Geologic and Hydrologic Sciences*. USGS Professional Paper 424-B: 136–39.

Schulz, W. H. 2007. *The Slumgullion Landslide, Hinsdale County, Colorado*. USGS Field Trip: http://landslides.usgs.gov/docs/schulz/FieldTrip_C.pdf

Williams, R. A., and T. L. Pratt. 1996. Detection of the base of Slumgullion landslide, Colorado, by seismic reflection and refraction methods. In *The Slumgullion Earth Flow: A large-scale natural laboratory,* USGS Bulletin 2130, eds. D. J. Varnes and W. Z. Savage, p.77–84.

35. BOX CANYON

Blair, R. 1996. *Geology of the Western San Juan Mountains and a Tour of the San Juan Skyway, Southwestern Colorado*. Colorado Geological Survey Open-File Report 96-4, Field Trip No. 9.

Ewing, T. E. 2017. Laramide and Cenozoic structural and paleotopographic history of the Ouray area and the northwestern flank of the San Juan Mountains, Colorado. In *The Geology of the Ouray-Silverton Area,* New Mexico Geological Society 68th Annual Fall Field Conference Guidebook, eds. K. E. Karlstrom and others, p. 169–78.

Gonzales, D. A., and S. Cumella. 2019. *Geology along the Ouray Area, June 1–2, 2019*. Available online at www.fourcornersgeologicalsociety.org.

Jeon, K., J. R. Giardino, and B. K. M. Gonzalez. 2017. Geomorphic characteristics of fens in the San Juan Mountains, Colorado. In *The Geology of the Ouray-Silverton Area,* New Mexico Geological Society 68th Annual Fall Field Conference Guidebook, eds. K. E. Karlstrom and others, p.187–94.

Johnson, B., M. Gillam, and J. Beeton. 2017. Glaciations of the San Juan Mountains: A review of work since Atwood and Mather. In *The Geology of the Ouray-Silverton Area,* New Mexico Geological Society 68th Annual Fall Field Conference Guidebook, eds. K. E. Karlstrom and others, p. 195–204.

36. SILVERTON

Cross, W., and E. S. Larsen. 1935. A Brief Review of the Geology of the San Juan Region of Southwestern Colorado. USGS Bulletin 843.

Garcia, R. V. 2011. *Cenozoic intrusive and exhumation history of the West Elk Mountains, southwestern Colorado*. MS Thesis, New Mexico Institute of Mining and Technology.

Lipman, P. W., and O. Bachmann. 2015. Ignimbrites to batholiths: Integrating perspectives from geological, geophysical, and geochronological data. *Geosphere* 11 (3): 705–743.

Lipman, P. W. 2006. *Geologic Map of the Central San Juan Caldera Cluster, Southwestern Colorado*. USGS Geologic Investigations Series I-2799.

Luedke, R. G., and W. S. Burbank. 2000. *Geologic Map of the Silverton and Howardsville Quadrangles, Southwestern Colorado*. Geologic Investigations Series Map I-2681, 1:24,000.

Parker, B. H. 1968. Placer gold in southwestern Colorado. In *San Juan, San Miguel, La Plata Region (New Mexico and Colorado)*, eds. J. W. Shomaker, New Mexico Geological Society 19th Annual Fall Field Conference Guidebook, p. 168–84.

37. LIZARD HEAD

Blair, R. 1996. Geology of the Western San Juan Mountains and a Tour of the San Juan Skyway, Southwestern Colorado. *Geologic Excursions to the Rocky Mountains and*

Beyond, Fieldtrip Guidebook for the GSA Annual Meeting, Oct. 29–31, 1996, GSA Special Publication 44.

Gonzales, D. A. 2015. New U-Pb Zircon and 40Ar/39Ar age constraints on the Late Mesozoic to Cenozoic plutonic record in the western San Juan Mountains. *The Mountain Geologist* 52 (2): 5–42.

Purington, C. W. 1898. Preliminary report on the mining industries of the Telluride quadrangle, Colorado. *USGS 18th Annual Report*, part 3, p. 745–850.

Raby, A. G., and J. S. Dersch. 1997. *Mineral Resource Potential and Geology of the San Juan National Forest, Colorado*. USGS Bulletin 2127.

38. ENGINEER MOUNTAIN

Blair, R. 1996. Geology of the Western San Juan Mountains and a Tour of the San Juan Skyway, Southwestern Colorado. *Geologic Excursions to the Rocky Mountains and Beyond*, Fieldtrip Guidebook for the GSA Annual Meeting, Oct. 29–31, 1996, GSA Special Publication 44.

Gonzales, D. A. 2015. New U-Pb Zircon and 40Ar/39Ar age constraints on the Late Mesozoic to Cenozoic plutonic record in the western San Juan Mountains. *The Mountain Geologist* 52 (2): 5–42.

Gonzales, D. A., and W. R. Van Schmus. 2007. Proterozoic history and crustal evolution in southwestern Colorado: Insight from U/Pb and Sm/Nd data. *Precambrian Research* 154 (1-2): 31–70.

Lipman, P. W., F. S. Fisher, H. H. Mehnert, and others. 1976. Multiple ages of Mid-Tertiary mineralization and alteration in the western San Juan Mountains, Colorado. *Economic Geology* 71 (3): 571–88.

Western Canyons and Plateaus

39. ECHO PARK IN DINOSAUR NATIONAL MONUMENT

Dehler, C. M., C. M. Fanning, P. K. Link, E. M. Kingsbury, and D. Rybczynski. 2010. Maximum depositional age and provenance of the Uinta Mountain Group and Big Cottonwood Formation, northern Utah: Paleogeography

of rifting western Laurentia. *GSA Bulletin* 122 (9-10): 1686–99.

Foos, A., and J. Hannibal. 1999. *Geology of Dinosaur National Monument*. National Park Service.

Hanson, W. R. 1983. *Geologic Map of Dinosaur National Monument and Vicinity*. USGS Map I-1407.

National Park Service. 2006. Dinosaur National Monument: Geologic Resource Evaluation Report. Natural Resource Report 2006/008.

Tressler, C. 2011. *From Hillslopes to Canyons, Studies of Erosion at Differing Time and Spatial Scales Within the Colorado River Drainage*. MS Thesis, Utah State University.

40 AND 41. TRAIL THROUGH TIME AND COLORADO NATIONAL MONUMENT

Aslan, A., K. E. Karlstrom, W. C. Hood, and others. 2008. River incision histories of the Black Canyon of the Gunnison and Unaweep Canyon: Interplay between late Cenozoic tectonism, climate change, and drainage integration in the western Rocky Mountains. In *Roaming the Rocky Mountains and Environs: Geological Field Trips*, GSA Field Guide 10, ed. R. G. Raynolds, p. 175–202.

Lohman, S. W. 1981. *The Geologic Story of Colorado National Monument*. USGS Bulletin 1508.

Scott, R. B., A. E. Harding, W. C. Hood, and others. 2001. *Geologic Map of the Colorado National Monument and adjacent areas, Mesa County, Colorado*. USGS Geologic Investigations Series Map I-2740, scale 1:24000.

42. BOOK CLIFFS

Cather, S. M., C. E. Chapin, and S. A. Kelley. 2012. Diachronous episodes of Cenozoic erosion in southwestern North America and their relationship to surface uplift, paleoclimate, paleodrainage, and paleoaltimetry. *Geosphere* 8 (6): 1177–1206.

Hood, K., and D. A. Yurewicz. 2008. Assessing the Mesaverde basin center gas play, Piceance Basin,

Colorado. In *Understanding, Exploring, and Developing Tight-gas Sands*, AAPG Hedberg Series 3, p. 87–104.

Lazear, G., K. E. Karlstrom, A. Aslan, and S. Kelly. 2013. Denudation and flexural isostatic response of the Colorado Plateau and southern Rocky Mountains region since 10 Ma. *Geosphere* 9 (4): 792–814.

Pederson, J. L., R. D. Mackley, and J. L. Eddleman. 2002. Colorado Plateau uplift and erosion evaluated using GIS. *GSA Today* 12 (8): 4–10.

43. GRAND MESA

Aslan, A., W. C. Hood, K. E. Karlstrom, and others. 2014. Abandonment of Unaweep Canyon (1.4-0.8 Ma), western Colorado: Effects of stream capture and anomalously rapid Pleistocene river incision. *Geosphere* 10 (3): 428–46.

Aslan, A., K. E. Karlstrom, E. Kirby, and others. 2019. Resolving time-space histories of Late Cenozoic bedrock incision along the Upper Colorado River, USA. *Geomorphology* 347: 106855.

Ellis, M. S., and V. Gabaldo. 1989. *Geologic Map and Cross Sections of Parts of the Grand Junction and Delta 30′ x 60′ Quadrangles, West-central Colorado*. Coal Investigations Map C-124, 1:100,000.

Karlstrom, K. E., D. Coblentz, K. Dueker, and others. 2012. Mantle-driven dynamic uplift of the Rocky Mountains and Colorado Plateau and its surface response: Toward a unified hypothesis. *Lithosphere* 4 (1): 3–22.

44. UNAWEEP CANYON

Aslan, A., W. C. Hood, K. E. Karlstrom, and others. 2014. Abandonment of Unaweep Canyon (1.4–0.8 Ma), western Colorado: Effects of stream capture and anomalously rapid Pleistocene river incision. *Geosphere* 10 (3): 428–46.

Aslan, A., K. E. Karlstrom, L. J. Crossey, and others. 2010. Late Cenozoic evolution of the Colorado Rockies: Evidence for Neogene uplift and drainage integration. In

Through the Generations: Geologic and Anthropogenic Field Excursions in the Rocky Mountains from Modern to Ancient, GSA Field Guide 18, eds. L. A. Morgan and S. L. Quane, p. 21–54.

Aslan, A., K. E. Karlstrom, W. C. Hood, and others. 2008. River incision histories of the Black Canyon of the Gunnison and Unaweep Canyon: Interplay between late Cenozoic tectonism, climate change, and drainage integration in the western Rocky Mountains. In *Roaming the Rocky Mountains and Environs: Geological Field Trips*, GSA Field Guide 10, ed. R. G. Raynolds, p. 175–202.

Donahue, M. S., K. E. Karlstrom, A. Aslan, and others. 2013. Incision history of the Black Canyon of Gunnison, Colorado, over the past ~1 Ma inferred from dating of fluvial gravel deposits. *Geosphere* 9 (4): 815–26.

Lohman, S. W. 1961. *Abandonment of Unaweep Canyon, Mesa County, Colorado, by capture of the Colorado and Gunnison Rivers*. USGS Professional Paper 424-B, p. B144–B146.

Oesleby, T. W. 1983 Geophysical measurement of valley fill thickness Unaweep Canyon, west central Colorado. In *Northern Paradox Basin-Uncompahgre Uplift: Grand Junction Geological Society 1983 Field Trip Guidebook*, ed. W. Averett, p. 71–72.

45. BLACK CANYON OF THE GUNNISON

Aslan, A., K. E. Karlstrom, W. C. Hood, and others. 2008. River incision histories of the Black Canyon of the Gunnison and Unaweep Canyon: Interplay between late Cenozoic tectonism, climate change, and drainage integration in the western Rocky Mountains. In *Roaming the Rocky Mountains and Environs: Geological Field Trips*, GSA Field Guide 10, ed. R. G. Raynolds, p. 175–202.

Donahue, M. S., K. E. Karlstrom, A. Aslan, and others. 2013. Incision history of the Black Canyon of Gunnison, Colorado, over the past ~1 Ma inferred from dating of fluvial gravel deposits. *Geosphere* 9 (4): 815–26.

Donahue, M. S., K. E. Karlstrom, K. E. Kelley, and J. W. Ricketts. 2013. Multi-stage uplift of the Rocky Mountains:

Using thermochronology data to unravel mechanisms and discrete episodes of uplift and intervening tectonic quiescence. *GSA Abstracts with Programs* 45 (7): 635.

Hansen, W. R. 1965. *The Black Canyon of the Gunnison: Today and Yesterday*. USGS Bulletin Report B: 1191.

Karlstrom, K. E., S. J. Whitmeyer, K. Dueker, and others. 2005. Synthesis of results from the CD-ROM experiment: 4-D image of the lithosphere beneath the Rocky Mountains and implications for understanding the evolution of continental lithosphere. In *The Rocky Mountain Region—An Evolving Lithosphere: Tectonics, Geochemistry, and Geophysics*, American Geophysical Union Geophysical Monograph 154, eds. K. E. Karlstrom and G. R. Keller, p. 421–41.

47. FOUR CORNERS

Armstrong, A. K., and L. D. Holcomb. 1989. Stratigraphy, Facies, and Paleotectonic History of Mississippian Rocks in the San Juan Basin of Northwestern New Mexico and Adjacent Areas. In *Evolution of Sedimentary Basins, San Juan Basin*. USGS Bulletin 1808-B-D.

Cather, S. M., M. T. Heizler, and T. E. Williamson. 2019. Laramide fluvial evolution of the San Juan Basin, New Mexico and Colorado: Paleocurrent and detrital-sanidine age constraints from the Paleocene Nacimiento and Animas formations. *Geosphere* 15 (5): 1641–64.

Dubiel, R. F. 2013. Geology, sequence stratigraphy, and oil and gas assessment of the Lewis Shale Total Petroleum System, San Juan Basin, New Mexico and Colorado. In *Total Petroleum Systems and Geologic Assessment of Undiscovered Oil and Gas Resources in the San Juan Basin Province, Exclusive of Paleozoic Rocks, New Mexico and Colorado*. USGS Digital Data Series 69–F, p. 1–45.

Fassett, J. E. 1974. Cretaceous and Tertiary rocks of the eastern San Juan Basin, New Mexico and Colorado. In *Ghost Ranch*, New Mexico Geological Society 25th Annual Fall Field Conference Guidebook, eds. C. T. Siemers, L. A. Woodward, and J. F. Callender, p. 225–30.

Gonzales, D. A. 2010. The enigmatic Late Cretaceous McDermott Formation. In *Geology of the Four Corners Country*, New Mexico Geological Society Fall Field Conference Guidebook 61, eds. J. E. Fassett, K. E. Zeigler, and V. W. Lueth, p. 157–62.

Lucas, S. G., and T. E. Williamson. 1992. Fossil mammals and the early Eocene age of the San Jose Formation, San Juan Basin, New Mexico. In *San Juan Basin IV: New Mexico Geological Society Fall Field Conference Guidebook*, eds. S. G. Lucas, B. S. Kues, T. E. Williamson, and A. P. Hunt, p. 311–16.

Williamson, T. E., and S. G. Lucas. 1992. Stratigraphy and mammalian biostratigraphy of the Paleocene Nacimiento Formation, southern San Juan Basin. In *San Juan Basin IV: New Mexico Geological Society Fall Field Conference Guidebook*, eds. S. G. Lucas, B. S. Kues, T. E. Williamson, and A. P. Hunt, p. 265–96.

48. MESA VERDE NATIONAL PARK

Harrison, G. W., and others. 2019. *Mesa Verde National Park Paleontological Resources Inventory (Non-Sensitive Version)*. National Park Service Natural Resource Report. NPS/MEVE/NRR—2017/1550.

INDEX

Grand Junction, 97, 100, 101, 102
Grand Lake, 18
Grand Mesa, 102–3
Grand Valley, 97, 98, 99, 105
granite, 6, 11, 16, 17, 38, 39, 50, 60; eroded pebbles of, 20, 28, 30; metamorphism from, 56; ore in, 51; Proterozoic, 16, 17, 60, 97, 114. See also Bakers Bridge Granite; Pikes Peak Granite; Silver Plume Granite
granodiorite, 46, 70
gravels, 9, 12, 102, 103, 104
Great Plains, 1, 9, 24, 46
Great Sand Dunes National Park, 76–77
Great Unconformity, vi, 3, 4, 20, 21, 24, 25, 37, 62, 63, 84, 85, 97, 114, 115
Green River, 94, 95
Green River Formation, 1, 7, 66, 101, 103
Grottos, 60–61
Grottos pluton, 60
groundwater, 9, 12, 36, 55, 64, 65, 66, 75, 78
Guffey volcanic complex, 40
Gunnison, 1, 2, 7
Gunnison River, 2, 69, 82, 97, 104, 105, 106, 107
gypsum, 15, 33, 65, 83

halite, 65
Hanging Flume, 108, 109
Hanging Lake, 62, 63, 64
Hanging Lake Tunnels, 62, 64
Harding Sandstone, 4
hematite, 20
Hermosa Group, 5, 85
High Plains Aquifer, 12
hogbacks, 14, 22, 23, 24, 26, 27, 33, 66, 68, 100, 101

Holocene, vi
hoodoos, 34
Hortense Hot Spring, 55
hot springs, 9, 49, 54, 55, 65, 79, 85, 114
hydrothermal, 55, 71, 82, 87

I-70 Roadcut, 22–23, 24
Idaho Springs Formation, 20, 42
Ignacio Quartzite, 114, 115
iguanodons, 24
incised, 82, 93, 109
inverted topography, 28, 102
iron oxides, 20

jarosite, 83
joints, 114
Jurassic, vi, 1, 6, 24, 33, 44, 94, 96, 98, 110, 111

Kayenta Formation, 109
Kelsey Lake Diamond Mine, 4
Kendall Mountain, 87
Keyhole arch, 14, 15
kimberlite, 4
Kirtland Shale, 111
K-Pg boundary, vi, 45

laccoliths, 70, 71
La Garita caldera, 80, 81
lahars, 40, 69
lakebeds, 76
landslides, 82, 102
Laramide orogeny, vi, 6–8, 11, 15, 16, 21, 36, 42, 49, 68, 97, 100, 106, 111
Laramie Formation, 7, 34
laumontite, 55
lava flows, 8, 73, 87, 102
lead, 50, 52, 53, 58, 86
Leadville, 50, 51, 52, 53

Leadville Limestone, 4, 52, 56, 62, 64, 65, 66, 84
Leadville mining district, 52–53
Lewis Shale, 111, 116, 117
limestone, 4, 5, 44, 52, 58, 59; caves in, 36, 37, 65; Leadville, 56, 62, 63, 64, 65, 66, 84; Manitou, 36, 37; metamorphosed, 56. See also travertine
Lizard Head, 88
Longs Peak, 3, 16–17
Longs Peak–St. Vrain batholith, 16
Loveland, 14, 15
Lykins Formation, 15, 22, 26, 32, 33
Lyons Formation, vi, 5, 22, 26, 27, 31, 32, 33
Lytle Formation, 22

Mancos Shale, 6, 7, 71, 78, 88, 93, 98, 100, 101, 103, 104, 106, 112, 113
manganese, 52, 113
Manitou Limestone, 4, 36, 37, 52
Manitou Springs, 3, 38
marble, 56, 57
Marble (town), 56, 57
Maroon Bells, 58–59
Maroon Formation, 5, 58, 59
Maroon Lake, 58, 59
McInnis Canyons National Conservation Area, 93
meandering, 109
Medano Creek, 77
Menefee Formation, 112, 113
Mesaverde Group, 7, 66, 68, 71, 98, 100, 101, 103, 111, 112
Mesa Verde National Park, 112–13
metamorphic rock, 16, 18, 42, 43, 106. See also gneiss; marble; quartzite; schist
metamorphism, 3, 49, 56, 58, 59, 71, 106

methane, 100, 111
microtektite, 45
Million Dollar Highway, 85, 89
Milner Pass, 18
Mineral Belt Trail, 53
mineralization, 71
minerals, 4, 6, 49, 50, 51, 52, 53, 87; deposited by water, 65, 76, 78, 82, 83. *See also specific mineral names*
mining, 4, 7, 50, 52, 53, 56, 58, 86, 87, 108. *See also* copper; gold; lead; molybdenum; ore; silver; uranium; zinc
Minturn Formation, 5
Miocene, vi, 12, 73, 103
Mississippian, vi, 4, 65, 66, 84, 111
Moenkopi Formation, 6
molybdenite, 50, 51
molybdenum, 50, 51, 71
Monte Vista National Wildlife Refuge, 75
Montrose Placer Mining Company, 108
monzonite, 54, 70, 106
moraines, 9, 17, 18, 88
Morrison, 24
Morrison Quarry No. 5, 24
Morrison Formation, 1, 6, 22, 23, 24, 26, 32, 33, 44, 94, 96, 98, 111
Mt. Princeton batholith, 54
Mt. Princeton caldera, 28
Mt. Princeton hot springs, 54
mudflows, 40, 69
mudstone, 28, 30, 33, 34, 40, 45, 58, 98, 113
muscovite, 106
Mygatt-Moore dinosaur quarry, 96

Nacimiento, 111
natural gas, 7, 11, 100, 101, 110
Naturita, 108
Niobrara, 11, 26, 32

North Gateway Rock, 32, 33

Ogallala Aquifer, 12
Ogallala Formation, 12, 13
oil, 7, 11, 100, 110, 111
oil shale, 7, 100
Oligocene, vi, 7, 8, 9, 12, 49, 50, 71, 73, 74, 80, 82, 88, 106
opal, 78
Ordovician, vi, 4, 36, 84
ore, 4, 6, 7, 8, 49, 51, 52, 53, 58, 71, 73, 87; *See also* copper; gold; lead; molybdenum; silver; uranium; zinc
oreodonts, 12
Oro City, 53
Ouray, 84, 85
Ouray fault, 84, 85
Ouray Limestone, 84, 85
outwash, 70

Pagoda Mountain, 9
Pagosa Springs, 78–79, 116
Painted Wall, 106, 107
Paint Mines Interpretive Park, 34–35
Paleocene, vi, 100, 111
Paradox Basin, 5
Paradox Evaporite, 5
Paradox Valley, 108, 109
patterned ground, 19
Pawnee Buttes, 12–13
Pawnee National Grasslands, 12
pegmatites, 105, 106, 107
Pennsylvanian, vi, 4, 20, 24, 26, 31, 58, 90, 95, 97
perched aquifers, 75
perimineralization, 41
Permian, vi, 5, 7, 20, 24, 26, 32, 58, 111
petrified wood, 1, 41
petroleum, 7, 93, 110
Piceance Basin, 7, 100, 101
Picketwire Canyonlands, 44

Pictured Cliffs Sandstone, 111, 116, 117
piedmont, 12
Piedra River, 116
Pierre Shale, 6, 7, 11, 26, 32, 34, 47
Pikes Peak, 3, 38–39
Pikes Peak batholith, 34, 38
Pikes Peak Granite, 3, 30, 37, 38, 39
Pinedale glaciation, 18
Pinkerton Hot Springs, 114
plagioclase, 28, 46
Platte River, 2
playas, 5
Pleistocene, vi, 9, 18, 42, 62, 74, 88, 89, 107
Pliocene, vi, 74
plutons, 6, 38, 60
Point Lookout Sandstone, 112
pollen, 41, 45
potholes, 61
Princeton, Mt., 28, 54, 55
Proterozoic, vi, 2–4, 6, 7, 18, 49, 61, 99, 105, 107; and Great Unconformity, 24, 25, 63, 84, 85, 97
Pueblo, 11, 42, 44
Purgatoire River, 44
pyrite, 6, 50, 52, 53, 82, 87

quartz: in granitic rock, 17, 38, 43, 46, 54, 70, 106; inclusions of, 56; in sandstone, 20, 30, 34; shocked, 45; in veins, 51. *See also* chalcedony; silica
quartzite, 28, 59, 84, 114, 115
quartz monzonite, 54, 70, 106
quartz syenite, 46

Rabbit Valley, 96
radial dikes, 46
radiometric, 7
Ralston Creek Formation, vi, 22, 24
Rampart Range fault, 31

Magdalena Sandoval Donahue grew up in northern New Mexico, fascinated by the mountains and valleys of the high desert at the southern end of the Rocky Mountains. She received a BS in Geological Sciences and a BS in Fine Arts from the University of Oregon, where she took structural geology from Marli Miller, her coauthor on this book. She obtained her MS and PhD from the University of New Mexico, and much of her research focused on the evolution of topography and mountain ranges in Colorado. She lives in Albuquerque with her husband, John, and three children.

Marli B. Miller is a senior instructor and researcher at the University of Oregon. She completed her BA in geology at Colorado College in 1982 and her MS and PhD in structural geology at the University of Washington in 1987 and 1992, respectively. Marli teaches a variety of courses, including introductory geology, structural geology, field geology, and geophotography. In addition to numerous technical papers, she is the author of *Oregon Rocks, Roadside Geology of Oregon, Roadside Geology of Washington* with coauthor Darrel Cowan, and *Geology of Death Valley National Park* with coauthor Lauren A. Wright. Because she maintained an active interest in Colorado geology since college, she was thrilled to become a part of this project. Marli has two daughters, Lindsay and Megan.